The Taconide Zone
and the Taconic Orogeny in the Western Part of the Northern Appalachian Orogen

E-an Zen

United States Geological Survey

Washington, D. C.

THE GEOLOGICAL SOCIETY OF AMERICA

SPECIAL PAPER 135

The printing of this volume
is made possible
through the bequest of
Richard Alexander Fullerton Penrose, Jr.

Library of Congress Catalog Card Number 70-184513
I.S.B.N. 08137-2135-0

Published by
THE GEOLOGICAL SOCIETY OF AMERICA, INC.
3300 Penrose Place
Boulder, Colorado 80301

Printed in the United States of America

To E-su

and the memory of

Shu-ming

Contents

Figure

Acknowledgments

This paper represents the cumulative results of occasional cogitations of many years, based in part on my own observations, often on field trips guided by leaders who generously shared their information and thoughts with me, and in part on interpretations or reinterpretations of published information. Although the paper no doubt includes a large element of personal bias in interpretation, I offer no apologies for this; one aim of the paper is to arouse discussion and stimulate fresh research in critical areas.

I do, however, apologize for any incorrect statement of facts. Many of these errors have already been weeded out by reviewers of various drafts of the paper. For their penetrating questions, instructive discussions, and patient and often extensive review both of the substance and style of the paper, I am deeply grateful to John M. Bird, William R. Church, Jack B. Epstein, Jarvis B. Hadley, Norman L. Hatch, Jr., Marshall Kay, Allison R. Palmer, William H. Poole, John Rodgers, Robert K. Stevens, and Walter S. White. If there be any credit to this paper, they are largely responsible.

For fruitful discussion of Appalachian geology over many years, I am grateful to these people, as well as to many others. It would be impossible to list all those who have contributed to my education in Appalachian geology, but I must at least acknowledge the following: W.B.N. Berry, W. D. Brückner, W. M. Cady, T. H. Clark, W. H. Diment, A. A. Drake, Jr., L. M. Hall, Yngvar Isachsen, P. H. Osberg, L. B. Platt, D. W. Rankin, N. M. Ratcliffe, J. C. Reed, Jr., John Riva, J. L. Rosenfeld, R. S. Stanley, J. B. Thompson, Jr., D. U. Wise, and Isidore Zietz.

Abstract

The lower Paleozoic rocks that extend from northwestern Newfoundland, through the Gaspé Peninsula, the south shore of the St. Lawrence River as far west as Quebec City, the Champlain Valley, western New England, eastern New York, and north-central New Jersey to southeastern Pennsylvania were deformed markedly by the Ordovician Taconic orogeny. This belt is bordered to the north and west by the little-disturbed foreland; in Canada the boundary includes Logan's Line. To the south and east, the identity of this belt is lost in rocks that have been more severely deformed and metamorphosed by later, principally Acadian, orogeny. Rocks of this identifiable Taconic orogenic belt are here termed the Taconides.

During Cambrian and much of Early Ordovician time, sedimentation within the Taconide belt was arranged in parallel zones: to the northwest, on the craton, was a shelf environment of shallow subtidal, intertidal, and supratidal carbonate deposition. Southeast of this shelf across a steep slope and an abrupt facies change, that probably reflects a sharp increase in water depth, was an area of clastic sedimentation; between these two zones was a zone of carbonate-clast slump conglomerates. The zone of clastic sediments, called the transitional zone, in turn passed seaward into one of typical eugeosynclinal sedimentation where the rocks are poor in carbonate, but rich in volcanic components.

Another major facies change is preserved within the area of the shelf sequence deposition. Depending on the geographic location, this second facies change is late Early Ordovician to Middle Ordovician in age; it is marked by the regional unconformable overlap of a syntectonic black-mantling shale sequence on the carbonate rocks of the shelf sequence and older units.

The shelf and mantling shale sequences are divisible into two tectonic zones: a foreland to the northwest, and a zone of deformed autochthonous rocks to the southeast, in which the intensity of deformation and metamorphism increases to the southeast. Rocks of the transitional sedimentary facies are even more intensely deformed, and over wide areas have been moved bodily northward and westward for long distances. These moved rocks are preserved in three forms: (1) klippe, now surrounded entirely by rocks of the shelf or mantling shale sequence across structural contacts; (2) allochthonous rocks only partly surrounded by rocks of the shelf or mantling shale sequence across contacts; (3) allochthons now eroded to a mere structural stump, but whose former extension and large movement are recorded by distinctive syntectonic sedimentary rocks. To the first category belong the Hare Bay and Humber Arm klippen of Newfoundland, small klippen in the Quebec City area, the Taconic klippe in New England and New York, and small scattered areas of allochthonous sedimentary rocks as well as a large klippe of Precambrian crystalline rocks in New Jersey and Pennsylvania. To the second category belong the rocks immediately southeast of Logan's Line, from the tip of Gaspé Peninsula to the Vermont-Quebec border. To the third category belong the rocks of the Hinesburg thrust in northern Vermont, where large-scale Ordovician movement at the surface level is recorded in the wildflysch-type sedimentary rocks exposed along the Lake Champlain shore.

South and east of the zone of allochthons and of the deformed shelf sequence is a zone of structurally high ground, part of the axial region of the composite Berkshire—Green Mountain—Sutton Mountain—Notre Dame Mountain—Shickshock Range—Indian Head Range—northern Long Range anticlinoria. Where the structural relief is especially great or where epeirogenic uplift has caused erosion to reach sufficient depth, the zone is marked by Precambrian basement rocks; these Precambrian rocks approximately mark the southeast limit of basement rocks of 1 (±) b.y. (billion years) age and appear to be the edge of a lower Paleozoic craton.

Coincident with the zone of structural highs is a zone of Bouguer gravity highs; the coincidence extends from northwestern Newfoundland through the Gulf of St. Lawrence, as far south as the north end of the Berkshire massif. From here south to Long Island Sound, the gravity ridge is displaced east of the structural ridge. It passes through the Coastal Plain deposits and reappears in the area of the Glenarm Series in Maryland. Northwest of the gravity ridge is a coextensive zone of Bouguer gravity troughs. The troughs follow the zone of deformed shelf sequence; where the allochthons occur, the troughs coincide with these features.

Lower to Middle Ordovician ultramafic rocks occur near the western boundary of the eugeosynclinal facies, in a narrow belt parallel with and just east of the zone of structural highs. From Gaspé Peninsula southwest, these ultramafic rocks are apparently strictly intrusive. In Newfoundland, however, intrusive ultramafic bodies may be genetically related to an apparently extrusive ultramafic-mafic ophiolite complex preserved in the allochthons.

I suggest that the process leading to the locations of the structural, igneous, and gravity features was the interaction of an oceanic segment of the crust with the adjoining craton. The location of this junction of crustal segments originally determined the location of the sedimentary facies junction between the shelf and basin sequences; compressive plunging of the oceanic crust under the craton caused rafting of the lighter cratonal margin, thus accounting for the structural uplift of the outermost (southeasternmost) known belt of 1-b.y.-old Precambrian basement rocks through much of the length of the Taconides. Farther into the craton (west and north), the compressive forces caused a gentle downwarp of the crust, leading to the subsidence of the former shelf area and, therefore, to a bathymetric reversal. The reversal allowed a black mantling shale sequence, whose sediments were derived in large part from the uplifted cratonal margin to the east, to be deposited over the former shelf area; continued uplift of the cratonal margin and subsidence of the former shelf area led eventually to wholesale emplacement of allochthons by gravity sliding of rocks of the transitional facies off the uplifted cratonal margin into the basin that was the former shelf area.

Continued compression in the Taconides after the initial submarine gravity sliding led to northwestward thrusting of consolidated rocks, including Precambrian crystalline rocks of the uplifted cratonal margin, in a more deep-seated environment, probably in Late Ordovician or Early Silurian time. Regional metamorphism accompanied this last stage of diastrophism.

The underthrusting of oceanic crust, in a process that probably involved the upper mantle as well, was accompanied by intrusion of the ultramafic bodies and the extrusion of siliceous, mafic, and ultramafic rocks on the surface. These igneous rocks are preserved today mainly in the eugeosynclinal sequence formerly deposited on the oceanic crust, but they are found also among the gravity slides and as volcanic ash in the shelf sequence.

The addition of a mass of relatively dense oceanic material under the margin of the craton, as well as the concomitant introduction of dense intrusive rocks, resulted in the belt of positive Bouguer gravity anomalies. Where the gravity ridge is southeast of rather than coinciding with the belt of Precambrian rocks in the Taconide zone, the Precambrian rocks have undergone large lateral transport toward the craton.

Compared to the Taconic orogeny, the Acadian orogeny in the northern Appalachian region was of wider regional extent, developed larger systems of nappes, led to more intense regional metamorphism, and was accompanied by larger scale plutonism. Despite these facts, however, the Taconic orogeny appears to have defined structural trends in the lower Paleozoic rocks that effectively controlled structural evolution of the northern Appalachian orogen during later Paleozoic orogenies, including the Acadian orogeny.

Introduction

Geologists studying the bedrock of the northern Appalachian region have long been aware of the existence of a line of major lithostratigraphic and structural discontinuity along the western front of the folded Appalachians. This is Logan's Line (for history, *see* Osborne, 1956, p. 168); it follows the St. Lawrence River from Gaspé Peninsula to Quebec City, thence south-southwest to the east shore of Lake Champlain; from here it trends almost due south to the upper end of that lake (Fig. 1). (Compass directions used in this paper are present-day directions. They are used for convenience of reference and do not necessarily indicate the correct early Paleozoic compass orientations. See Appendix 1.)

Recent geologic and geophysical studies show that Logan's Line is not simply a trace of a surface of discontinuity in the bedrock, but the composite western limit of a zone characterized by distinctive early Paleozoic sedimentary and tectonic history. This paper is an attempt to describe and summarize some of this geologic history.

The rocks of the Paleozoic Appalachian belt that mostly have been affected only by the Taconic orogeny will be called the Taconides. As the term is used in this paper, the Taconide belt is bordered on the north and west by Logan's Line where the line has been defined; otherwise, the northwest limit of the belt is the southeast limit of the foreland of the Taconic orogeny. On the southeast side of the Taconide belt, our discussion will be restricted to the area north and west of the Cabot fault zone in Newfoundland, the Shickshock Range on the Gaspé Peninsula, the axial region of the Notre Dame Mountain—Sutton Mountain anticlinorium in the Eastern Townships of Quebec, and the Hinesburg thrust in Vermont. The eastern limit of the Taconide belt then follows the Green Mountain and Berkshire massifs of Precambrian rocks (hereafter called the Precambrian massifs). The Taconide belt is some 100 km wide and over 2,000 km long; yet within this narrow belt many important Paleozoic geologic phenomena are persistent: a fundamental lateral change in sedimentary facies, the location of both detachment and receiving sites of major gravity slides and other thrust sheets, a persistent pair of large Bouguer gravity anomalies, and a possible change from continental to oceanic crust (which has since become altered by tectonic, plutonic, and metamorphic activities).

In the first section of this paper, we describe the geologic features that seem to characterize the Taconides. The second section describes individual geologic terranes; because of uneven amount of published work and the writer's own familiarity in these areas, the description is necessarily uneven in detail. The final section of the paper will attempt to interpret the origin and history of the Taconides.

Regional Relations

The west flank of the northern part of the Appalachian orogen can be divided into five narrow, but persistent, belts of rocks (Fig. 1). Each of these belts is characterized primarily by the nature and age of the stratigraphic sequence, but secondarily by its tectonic history. From northwest to southeast, the following belts may be distinguished:

Zone 1. The Foreland

Rocks of the foreland rest on the stable craton. Cambrian and Ordovician foreland rocks are found on both shores of the Strait of Belle Isle in Quebec and Newfoundland, in the St. Lawrence Lowland of southwestern Quebec, and in the Champlain, Hudson, and Mohawk Valleys in New England and New York. Between the areas of the Gulf of St. Lawrence and the St. Lawrence Lowland, Paleozoic rocks of the foreland are absent. The "foreland sequence" at Anticosti Island, in the St. Lawrence Valley as well as at nearby shorepoints are "foreland" with respect to later orogenies, but not necessarily with respect to the Taconic orogeny because the rocks are post-Taconic (*see* Append. 2, however, for a discussion of the drill-hole data from Anticosti Island).

South of the latitude of Albany, New York, the Cambrian and Ordovician foreland sequence disappears under younger rocks of the Catskill, Pocono, and Allegheny plateaus, and merges into the folded and thrust-faulted Valley and Ridge province. These foreland rocks are probably deformed by thin-skinned tectonics (Rodgers, 1963); the deformation is likely an Acadian or even Appalachian event, and during the Taconic orogeny the area could have been part of the true foreland.

Zone 2. Deformed Sedimentary Rocks

Deformed carbonate rocks and quartzite, stratigraphically continuous with those of the foreland and ranging from Lower Cambrian to Middle Ordovician, occur south and east of the foreland. The rocks are of the shelf facies; they have been folded, faulted, and metamorphosed — the degree of deformation and recrystallization increasing to the east and south. The rocks are autochthonous or at the most parautochthonous (*see* Append. 4 for definitions), having been transported relatively small distances. This belt is missing in the area between Gaspé Peninsula and approximately Quebec City — in much of the interval where the classical "Logan's Line" is recognized. This belt of rocks directly underlies much of western Newfoundland, including Belle Isle (Williams and Stevens, 1969); it underlies western Vermont, westernmost Massachusetts, northwestern Connecticut, the Dutchess County and New York City area of New York, part of north-central New Jersey, and southeastern Pennsylvania.

Zone 3. Allochthons

This zone is at places within zone 2, at places near the junction of zones 1 and 2, and at places taking the place of zone 2 altogether. Klippen are found in Newfoundland, Quebec, New England and New York, southeastern Pennsylvania and north-central New Jersey. Most of the allochthonous rocks south and southeast of Logan's Line in Quebec are not demonstrable klippen; they are paraklippen (*see* Append. 4 for definitions). Rocks of the allochthons are shales, arkoses, and graywackes, with subsidiary to significant amounts of mafic volcanic and ultramafic intrusive rocks as well as carbonate beds and conglomerates.

Zone 4. Precambrian Crystalline Rocks

This zone is east of zones 2 and 3, but apparently within the eastern part of zone 2 locally. From north to south, these are the massifs of the northern Long Range and Indian Head Range of western Newfoundland, the Green Mountain, Berkshire and Housatonic massifs of western New England, and the Hudson massif of New York, which continues into New Jersey and southeastern Pennsylvania as the Reading Prong. The structural relation of each massif to the surrounding Paleozoic rocks varies. The Green Mountain massif is almost certainly autochthonous, although likely, in part, fault-bounded on the west side. The Berkshire massif is bounded on the west side by major thrust faults and, thus, is parautochthonous. The Hudson and even the Housatonic massifs may prove to be allochthonous; the south end of the Hudson massif in New Jersey and Pennsylvania almost certainly is. The structural relation of the northern Long Range massif remains obscure, but it is partly fault-bounded on the west side. The autochthonous massifs form cores of major anticlinoria in the Appalachian orogenic system and have yielded isotopic ages showing that the rocks are at least as old as about 1 b.y. These crystalline rocks have defined important linear features since the beginning of the Paleozoic, for the Paleozoic rocks on the east side of this zone are of a different facies from those on the west side.

Zone 5. Lower Paleozoic Eugeosynclinal Rocks

East of the Precambrian crystalline rocks begins a broad terrane of probably autochthonous lower Paleozoic eugeosynclinal volcanic and sedimentary rocks, now metamorphosed to schist, impure quartzite, amphibolite, and granitic gneiss; they were originally shale, arkose, graywacke, basalt, rhyolite, and andesite. Compared with rocks of zone 3, these rocks contain abundant volcanic and few carbonate rocks and are vastly thicker. In addition, this zone contains a linear, persistent belt of ultramafic intrusions. These intrusions occur in Newfoundland just east of the area of Cabot fault; they occur on Gaspé Peninsula and continue southwestward, through the Thetford area into eastern Vermont, west-central Massachusetts and Connecticut, and Staten Island, and reappear from under Coastal Plain deposits to form the prominent belt of ultramafic rocks along the Pennsylvania-Maryland border. These ultramafic intrusions are probably of Early Ordovician age.

Because of later, probably Acadian, metamorphic and deformational effects, the Ordovician orogenic imprint is rarely observed in zone 5. The main features that attest to the Taconic effect are the ultramafic intrusions, intrusions of granitic rocks of about the same age (some rocks in the Oliverian gneiss domes and the Highlandcroft Plutonic Series; *see* Naylor, 1968), and local evidence of pre-Acadian deformational patterns in western New England (*see* for example Pavlides and others, 1968; Rosenfeld, 1968). At least the western part of this zone must belong to the Taconides; because of the lack of a clear-cut boundary in

the field, however, the southeastern limit of the Taconide zone is ambiguous.

Radiogenic dating of events in the Taconides is scattered and has uneven stratigraphic control. Many of the existing numbers are single-mineral, whole-rock, or mineral-isochron age determinations; their geologic interpretation varies and the numbers cannot be directly compared. Nevertheless, many dates obtained on regionally metamorphosed rocks suggest minimum ages of metamorphism and, thus, shed light on the geologic history. In the southwestern Eastern Townships, Quebec, the Taconide zone yields ages in the 440(±) m.y. range. To the southeast, ages of 350(±) m.y. begin to occur along the crest of the Green Mountain anticlinorial axis near the international boundary. In Vermont and adjacent parts of New York and Massachusetts, the 440(±) m.y. age applies west of the Green Mountain massif, but the 350(±) m.y. age applies to the east. The younger age "front" swings to the southwest in southwestern Massachusetts, so that 440(±) m.y. ages do not exist in Connecticut. The older dates are found again in a few determinations from the Manhattan Prong and from southeastern Pennsylvania. On the whole, it appears that the Taconide zone, recognized by its structural history, largely has escaped Acadian orogenic and younger clock-setting effects, though Acadian effects are found in the Taconides in southwestern Connecticut and nearby parts of New York and New Jersey (*see* Tilton and others, 1958, 1959; Long and Kulp, 1962; Rickard, 1965; Rodgers, 1967; Clarke and Kulp, 1968; Harper, 1968; Lyons and Faul, 1968; Ratcliffe, 1970; Wetherill and others, 1968, for primary and secondary sources of information).

Characterization of the Taconide Zone

Introduction

Despite the great diversity of geologic relations on the west flank of the northern Appalachian orogenic system, extending from Newfoundland to the Potomac River, there are, nevertheless, several large-scale geologic and geophysical features that persist and bear distinct spatial and possibly chronologic relations to one another. Because of their scale and because events leading to their formation can be dated to or associated with the early Paleozoic "Taconic" orogeny, these features are fundamental entities in the orogenic process and, as such, furnish clues to the reconstruction of the geologic history. The features that we will consider are two major sedimentary facies changes, a belt of ultramafic intrusive rocks, a zone of allochthons, a belt of Bouguer gravity lows coinciding with the zone of allochthons, a complementary belt of gravity highs, and finally, a zone of longitudinal high-angle faults that is particularly well displayed around the Taconic klippe.

Sedimentary Facies Changes

Two major changes in sedimentary facies exist in the Taconide zone. The first facies change is a lateral passage, in the direction away from the craton, from the carbonate-quartzite assemblage typical of the shelf to a predominantly shale, arkose, graywacke, and volcanic assemblage of deeper water, off-shelf, basin sequence. For simplicity, these two sequences will be referred to hereafter as the shelf and basin deposits, respectively, without necessarily implying the geometry of the depositional areas. In general, the shelf deposits are to the north and west of the basin deposits. The shelf deposits are largely carbonate rocks of organic origin deposited in supratidal, intertidal, and shallow subtidal, clear-water environments, where conditions were fairly uniform over large areas and apparently were also persistent with time. Details of origin of many of the carbonate rocks have been partly obliterated by recrystallization, especially in the folded and metamorphosed part of the sequence.

The basin sequence was deposited in an environment which was variable laterally as well as vertically (Cady, 1967) and apparently of deeper water than that of the shelf sequence. Two subsequences can be distinguished. The northwestern subsequence adjoins the shelf sequence, and consists of shale and graywacke, with subsidiary carbonate and volcanic rocks. This subsequence is relatively thin, probably a "starved" sequence (Elam, 1960), and has been interpreted as one deposited near the axial part of a trough (Zen, 1967a; *see also* Append. 3); sediments were supplied from both the northwest (craton) and the southeast (geosyncline). We will refer to this subsequence as the transitional facies. The second, southeast subsequence, when compared with the transitional facies, is much richer in volcanic rocks, poorer in carbonate rocks, and is much thicker — several times the thickness of the transitional facies or of the shelf sequence (Cady, 1968a). This subsequence will be called the eugeosynclinal facies.

The nature of the original sedimentary junction between the shelf sequence and the basin sequence varies from place to place. On the whole, this junction is fairly abrupt, and, though in a given outcrop it might appear as a normal lithic succession, the two facies seesaw forth and back so that the geographic location of the facies junction is constant to within about 10 km. Shale and sandstone, and locally chert, intercalate on outcrop scale with carbonate rocks, but mixing of the sedimentary components in individual beds appears minimal.

Within this zone of facies junction a peculiar kind of sedimentary rock is developed — carbonate-clast conglomerates that have a predominantly carbonate and quartz-sand matrix, but also a subsidiary shale matrix, the proportion of the latter increasing in the presumed offshore direction. In detail, these conglomerates fall into many different lithic classes: some are polychronous and even include clasts of crystalline basement rocks, but others are monochronous and are of the age of the enclosing bedded sedimentary rocks; some have giant, ill-sorted and poorly rounded clasts, but others have small, well-sorted and well-rounded clasts; some have virtually no matrix material, but others contain clasts floating in matrix. Despite this variability, these conglomerates help to decipher the original geographic and topographic arrangements of the sedimentary basin. The conglomerates are being considered in detail elsewhere (Zen and J. M. Bird, unpub. data) and will not be described further in this paper.

The facies transition just described extends from north-central Newfoundland to the Potomac River, and persisted from Early Cambrian to the early part of the Middle Ordovician (the precise time limits depending on the locality). Factors maintaining this transition were terminated by tectonic events to be described.

The second facies transition that covered an extensive area of the former shelf terminates this first facies transition in the geologic record. This second transition is an upward change of sedimentation: the carbonate and quartzite sequence of the shelf facies is succeeded upward by a black shale whose lower strata include carbonate beds or even carbonate-clast conglomerates (Zen and J. M. Bird, unpub. data). Over wide areas, this vertical facies change is defined by an unconformity, so that the black shale unit locally rests on rocks as old as the Precambrian; in adjacent areas, however, an unconformity may be absent and an apparently conformable succession is preserved. The base of this black shale unit ranges in age, depending on location, from latest Early Ordovician to late Middle Ordovician; however, the geologic role of the shale unit in the sedimentary-tectonic evolution of the Taconide zone seems everywhere the same. The black shale is the first definitive mark of diastrophism in its source area; I will refer to it as the "mantling shale unit."

Higher in the stratigraphic section, the black shale becomes progressively richer in sandstone, arkose, and graywacke; mafic and ultramafic detritus has been recorded in many places. The graywackes, recorded throughout the length of the Taconide belt (*see*, for instance, Behre, 1933; Osborne and Riva, 1966; Zen, 1967a; Tuke, 1968; Enos, 1969; and Drake, 1969, for descriptions and further references), are second-cycle sediments; they indicate a continued eastern sedimentary source as well as coarsening of sediment supply with passage of time. Scanty paleocurrent data (McBride, 1962; Middleton, 1965; Hawley, 1969; Enos, 1969, p. 48-53) on the graywacke are not all mutually consistent; McBride (1962) and Enos (1969) reported directions having easterly components. Middleton (1965) reported westerly components for the graywackes in the Albany, New York, area but the rocks are complexly deformed, and the uncertain interpretation of the effects of these deformations beclouds the issue (Zen, 1968, p. 136).

Allochthons and Related Structures

Large allochthons distinguish the Taconide zone. These allochthons have similar internal stratigraphy and also bear comparable stratigraphic relations to the surrounding autochtons (Fig. 2). The allochthons can be grouped, according to the rocks involved and mode of emplacement, into two general categories. The first and more important category contains klippen and paraklippen (*see* Append. 4) of lower Paleozoic sedimentary rocks that have been emplaced as submarine gravity slides. These slides are found in the Hare Bay and Humber Arm klippen in western Newfoundland; in the klippen in the Quebec City and Kamouraska area, 50 km northeast of Quebec City; in the Taconic klippe of western New England and nearby New York; and in isolated, discontinuous allochthonous rocks in north-central New Jersey and southeastern Pennsylvania. In addition, rocks immediately south and southeast of Logan's Line, extending from the tip of Gaspé Peninsula to near Granby, Quebec, are probably mostly if not all allochthonous, though much of the detailed structural and geometric relations there are imperfectly understood. Finally, other rock masses in the Taconide zone, by their lithic similarity to known allochthonous rocks and their anomalous geometric position, or by the presence of rocks that elsewhere serve as guides to allochthons, indicate possible allochthonous masses (the New York City area; Hall, 1968a, p. 13) or former presence of allochthons now eroded away (the St. Albans, Vermont, area).

These allochthons of sedimentary rocks are gravity slides; they consist predominantly of shale, sandstone, and graywacke and are interpreted as of a deep-water facies. Igneous rocks, mainly mafic and ultramafic rocks, are subsidiary. Carbonate rocks chiefly occur as carbonate-clast conglomerates of northern or western sedimentary derivation and are found mainly on the west and north sides of the allochthonous masses. The emplacement of the allochthons was from the south and east, into a basin of deposition of the mantling black shale over the former shelf area; wildflysch-type conglomerate is the ubiquitous harbinger of the imminent arrival of the slides. The association of black mantling shale, wildflysch-type conglomerate, and the gravity slides permits dating of the sliding events to various stages of late Early or Middle Ordovician and indicates a genetic relationship among the processes.

The second category of rocks, here considered to be allochthons, consists of both sedimentary and crystalline "basement" rocks; these allochthons bear no evidence of submarine sliding, and wildflysch-type conglomerate is conspicuously absent at their bases. These allochthonous rocks are the higher slices of the Taconic klippe and the probable klippe of Precambrian rocks of north-central New Jersey and southeastern Pennsylvania (the Reading Prong); the northeastern extension of this mass, the Hudson Highlands, may also be allochthonous. Emplacement of these masses was also from the south and east, probably during the latest Ordovician, younger than the episode of gravity sliding.

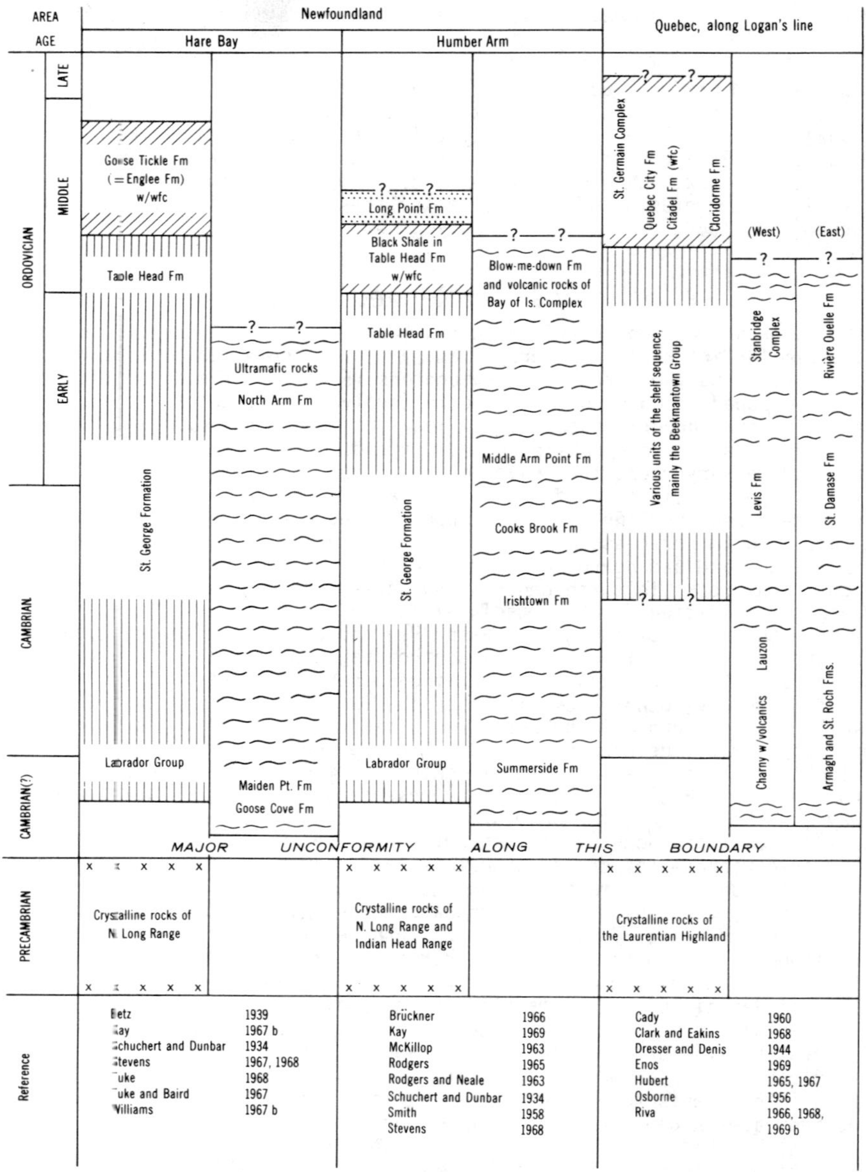

Figure 2. Comparison of stratigraphic sections in six major areas of the Taconides. In each area, the left-hand column shows the section in the autochthon, consisting of the basement, the shelf sequence and the mantling black shale, and the right-hand column shows the section in the allochthon, consisting of the basin sequence and locally of crystalline Precambrian rocks. Symbols: crosses, crystalline Precambrian rocks; vertical lines, rocks of the shelf sequence; diagonal lines,

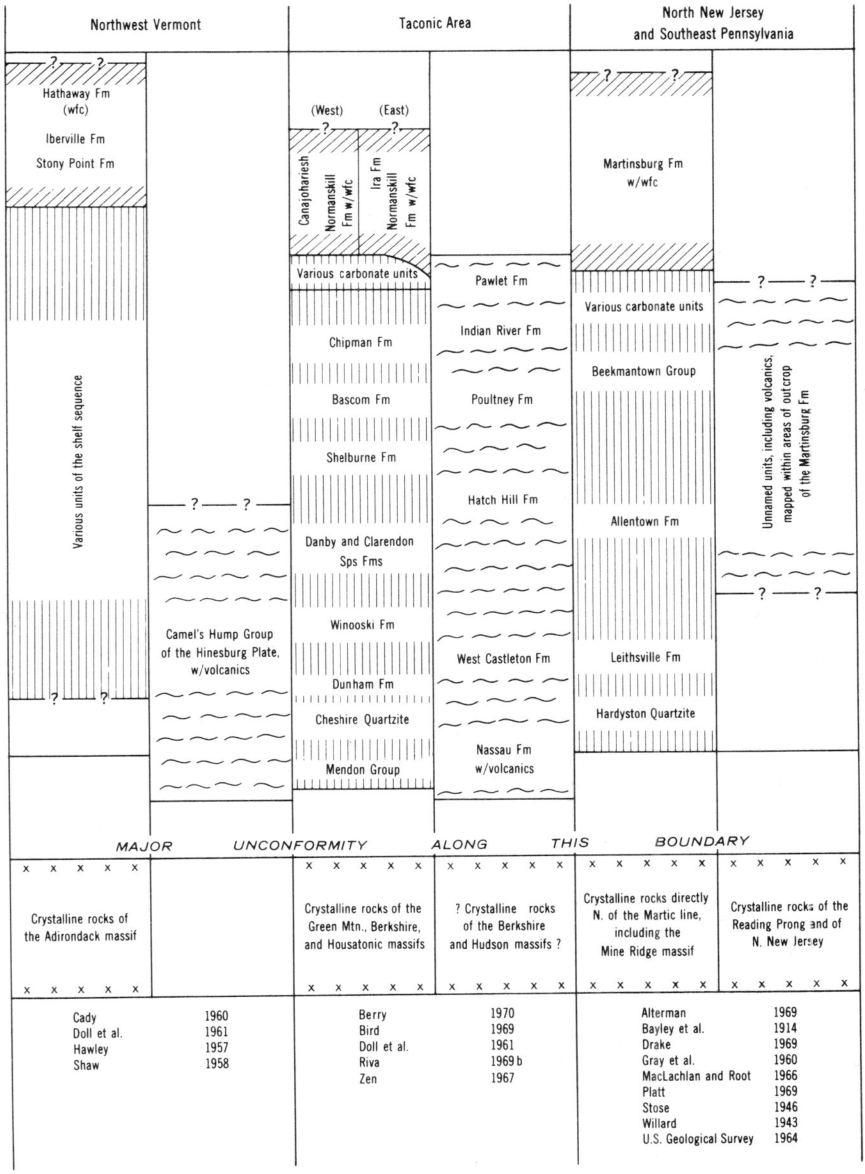

rocks of the syntectonic mantling shale sequence; wiggly lines, rocks of the basin sequence; dots, rocks that are post-tectonic with respect to the Taconic orogeny. Only major stratigraphic units are listed, and these are highly generalized. The abbreviation wfc stands for wild-flysch-type conglomerate. Principal references used in the compilation are listed for each area.

Longitudinal Faults

Longitudinal high-angle faults are developed in the lower Paleozoic autochthonous shelf area surrounding the Taconic klippe. These faults are approximately synchronous with the initial deposition of the mantling black shale sequence, and faulting may have been locally accompanied by volcanism (Thompson, 1967, p. 88). These faults probably were responsible for the fragmentation of the shelf area and its subsequent foundering; block faulting is considered an important mechanism for the bathymetric reversal needed for the initiation of gravity sliding.

In other parts of the Taconide zone, high-angle faults also exist. Data on these faults are scanty, and their genetic association with the Taconic orogeny cannot be established; however, by analogy with the relations established in the Taconic klippe area, we surmise that many high-angle faults found in the foundered shelf area surrounding other allochthons had a similar geologic history.

Igneous Rocks

The most important igneous rocks in the Taconide zone are ultramafic bodies. These are intrusive rocks in a restricted linear belt east of the zone of Precambrian massifs, from western Connecticut to the Gaspé Peninsula (Cady, 1960, 1968a; St-Julien, 1968; Chidester, 1968), and east of the belt of Carboniferous rocks in western Newfoundland (Fig. 6, *see* Rodgers and Neale, 1963; Poole, 1967). Some of the intrusions afford approximate stratigraphic dates; available data indicate that most of these intrusive rocks are Ordovician in age.

In the Eastern Townships, Quebec, ultramafic bodies intrude Cambrian and Lower Ordovician rocks (St-Julien, 1967); two K-Ar dates on muscovite from a granite and a pegmatite that intrude the ultramafic rocks in this belt yielded 477 m.y. and 481 m.y. ages (Leech and others, 1963, p. 78). In this same belt, pebbles of ultramafic rocks have been found in a Middle Ordovician unit (Magog Formation, St-Julien, 1961; *see* Berry, 1962; Riva, 1968) and in an Upper Ordovician conglomerate (Lamarche, 1962, p. 17; St-Julien, 1965). In north-central Vermont the intrusion is probably post-Early Ordovician; in south-central Vermont ultramafic bodies intrude Middle Ordovician rocks (*see* Doll and others, 1961; Currier and Jahns, 1941).

Enos (1969, p. 47-51, Table 1) found grains of serpentinite and of chromite in the late Middle Ordovician mantling black shale (Cloridorme Formation) along the northeast shore of Gaspé Peninsula (Fig. 1). Enos (p. 47) stated that the chromite is similar in properties to the marginal chromite of the Mt. Albert ultramafic intrusion, though paleocurrent data suggest that Mt. Albert itself was not the source of the chromite pebbles (p. 49). Two micas in the contact aureole of the Mt. Albert intrusive body have been isotopically dated at 495 m.y. (Lowden and others, 1963, p. 107).

In western Newfoundland, ultramafic bodies, partly possibly extrusive, are found in the Humber Arm and Hare Bay klippen. Although the igneous rocks are not isotopically dated, field evidence shows that a late Early to early Middle Ordovician age is probable: the intrusion was no older than Early Ordovician, but was older than the gravity sliding; consistent with this dating, detrital clasts of ultramafic rocks are found in autochthonous Middle Ordovician black shale in the Hare Bay area and in the Ordovician rocks of the Humber Arm klippe area (Stevens and others, 1969; Stevens, 1970, p. 170).

Ultramafic rocks of the Burlington Peninsula in Newfoundland (Fig.

6) are pre-Silurian, although some may be younger. Williams (1967a, p. 120), drawing upon previous work, showed that pebbles of the ultramafic rocks that intrude the Lower Ordovician Snooks Arm Group (Snelgrove, 1931, p. 485) are found in the Cape St. John Group, which has been correlated with known Silurian rocks. Ultramafic bodies in the Baie Verte Group (Church, 1969, Fig. 2) may be of Ordovician age or younger, depending on whether the Baie Verte Group is of Early Ordovician or Silurian age (*see* Neale and Kennedy, 1967, p. 163).

Note added in proof: On the Newfoundland ultramafic rocks, both in the Humber Arm klippe and on Burlington peninsula, see the following important recent contributions: 1. Kennedy, M. J., and W. E. Phillips, 1971, Ultramafic rocks of Burlington peninsula, Newfoundland: Geol. Assoc. Canada Proc., v. 24, p. 35-46. 2. Williams, Harold, 1971, Mafic-ultramafic complexes in western Newfoundland Appalachians and the evidence for their transportation: A review and interim report: Geol. Assoc. Canada Proc., v. 24, p. 9-25.

Other ultramafic intrusions in the same belt of the Taconides cannot be fixed in age. Such intrusions occur in Massachusetts (Emerson, 1917), Connecticut (Stanley, 1964; Gates and Christensen, 1965), near New York City (Staten Island — Broughton and others, 1962; and in the Hartland Formation — Hall, 1968b, p. D6-7) and in southeastern Pennsylvania where large numbers of serpentinite bodies occur (Gray and others, 1960). Field relations for these intrusive bodies are compatible with an Ordovician age, though not necessarily demanding it. One serpentinite body in southeastern Pennsylvania has a probable minimum age of 460 m.y. based on radiogenic dating of micas from a pegmatite that appears to intrude the margin of the serpentinite (Lapham and Bassett, 1964).

Although volcanic rocks are known in the northern Appalachian eugeosynclinal sequence beginning with the Cambrian, the Ordovician was a time of especially active volcanism. Mafic volcanic rocks dominate; examples are the Ammonoosuc Volcanics of the Connecticut Valley—Gaspé synclinorium (Thompson and others, 1968; Cady, 1960), the volcanic rocks of the Missisquoi Formation in Vermont (Doll and others, 1961), the Bolton volcanics in the Eastern Townships, Quebec (Cady, 1960), and extensive mafic and intermediate volcanic rocks in Newfoundland (*see*, for instance, Rodgers and Neale, 1963; Kay, 1966, 1967a; Poole, 1967; Neale and Kennedy, 1967; Horne, 1969). Intermediate to silicic volcanic rocks occur in the Missisquoi Formation and in the Ammonoosuc Volcanics; they probably also occur in the Bronson Hill anticlinorium as metamorphosed gneisses of the Oliverian Plutonic Series, perhaps recording a chain of volcanic islands in the eugeosyncline, to which the granitic intrusions of the Highlandcroft Plutonic Series are intimately related (Naylor, 1968; Thompson and others, 1968; Berry, 1968).

Volcanic ash deposits, including bentonite beds, are well known in the Middle Ordovician shelf sequence of the Appalachian region. In the northern Appalachian region, bentonite beds are recorded at scattered places in the circum-Adirondack region and in the Mohawk Valley in New York and in the Lake Champlain area in Vermont (Kay, 1935; *see also* Fisher, 1962), and ash beds have been reported from the unmetamorphosed, autochthonous Cloridorme Formation in northeastern Gaspé Peninsula (Enos, 1969, p. 29). Presumably the ash was derived from explosive volcanoes in the eugeosyncline.

Bouguer Gravity Anomalies

Enough Bouguer gravity surveys have been carried out in the northern Appalachians in recent years so that a broad survey of the gravity relations in the Taconides is possible (*see* Woollard and Joesting, 1964; Dominion Observatory Ottawa, 1967; Diment, 1968). Detailed relations for the individual areas are reviewed in the next section; here we summarize the principal features (Fig. 3).

A belt of gravity highs is paired in the Taconide zone with a belt of lows. The belt of lows occurs on the northwest side of the Taconide zone, roughly coincident with the present location of the belt of allochthons. The belt of highs occurs on the southeast side, coincident with the zone of Precambrian massifs as far south as the Berkshire massif, but is east of the Precambrian rocks farther southwest. The lows are more open features than the highs. Each belt of anomaly consists, in detail, of discrete and separate anomaly summits and troughs, but all these separate features are aligned in two narrow, defined, and spatially associated zones that extend from Newfoundland to Alabama.

The problem of origin of the gravity anomalies in this belt has been considered by many investigators (Diment, 1968; Weaver, 1967; Innes and Argun-Weston, 1967). Both Diment (1968) and Weaver (1967) recognized that the density differences among the exposed rocks cannot account for the anomalies.

Weaver (1967, p. 242) suggested that the negative anomaly in Newfoundland could be explained by a layer of sialic rock some 5 km thick and emphasized the existence of granitic rocks in the exposed Precambrian basement. However, areas of Precambrian granitic gneisses in western Newfoundland, as well as similar areas in New England, coincide with gravity highs, not lows; moreover, there is no geologic evidence for a belt of granitic batholiths along the belt of Bouguer gravity lows.

Both Diment (1968, p. 404) and Innes and Argun-Weston (1967) indicated that the gravity high can be explained by upwarp of uppermost mantle material, roughly wedge-shaped in cross section, that is some 20 km wide at the base and 20 km in relief. A structural relief of about 20 km presumably would be reflected by a large structural relief at the surface; this agrees with the geologic data along the western front of the chain of Precambrian massifs. The general picture of a wedge of dense material is supported by the interpretation of seismic data in the area around Newfoundland (Sheridan and Drake, 1968).

Discussion of Specific Areas

Northwestern Vermont

The original lower Paleozoic sedimentary facies relations can be reconstructed most readily in northwestern Vermont (Fig. 4). Thus, this area can serve as a template for the restoration of the more complex and allochthonous areas.

In northwestern Vermont, rocks of the shelf facies, now overlain by the Middle Ordovician mantling black shale sequence, occur along the shores of Lake Champlain as part of the little-deformed foreland (*see* Doll and others, 1961). The foreland is terminated to the east by the Highgate Springs thrust slice on which a more eastern phase of the shelf facies has moved westward. The Highgate Springs slice is succeeded to the east by the structurally complex Champlain slice (by the Philipsburg slice in Quebec; Fig. 9); rocks in the western part of the Champlain slice are largely carbonate, but contain noticeable calcareous shales; in the eastern part, shales predominate (Shaw, 1958). The gradation between the two facies is by lithic intercalation, by the presence of carbonate-clast conglomerates, and by changing proportions of shaly and calcareous components in a given unit (*see* Cady, 1960, 1968a). As a result, within the Champlain slice, the facies of the lower Paleozoic rocks changes eastward and locally northward from a facies typical of the shelf to one bearing many hallmarks of a eugeosyncline.

Rocks of the Champlain slice are truncated to the east by the Hinesburg thrust. Rocks above this thrust fault constitute the Hinesburg plate and consist largely of shale and graywacke, with intercalated mafic volcanic rocks, typical of the eugeosynclinal sequence. These rocks are Cambrian(?) and Lower Cambrian near the thrust fault. At the base of the section in the Hinesburg plate, immediately adjacent to the fault, the Cheshire Quartzite (Lower Cambrian, Doll and others, 1961) has been mapped; the Cheshire grades northward and eastward into a green phyllite in Quebec (Osberg, 1965). The Cheshire Quartzite is also identified in the Champlain slice. Thus, despite the interposition of numerous thrusts, some of uncertain stratigraphic throw, there appears to be a composite, but fairly complete record here of early Paleozoic east-west gradation between rocks of the shelf and basin facies (for more details, *see* Cady, 1968a).

The inception of the Hinesburg fault has been assigned (*see* discussions *by* Cady, 1960, p. 564; 1968b, p. 566; 1969, p. 64) to the Acadian orogeny. Sedimentary record, however, indicates a late Middle Ordovician age.

The mantling black shale in the foreland is as young as late Trenton (Middle Ordovician, Hawley, 1957, p. 58). Both on the east shore of Lake Champlain and on numerous islands, the uppermost part of this black shale contains a wildflysch-type polymictic conglomerate. The clasts include graywacke and chert, that, as far as I know, can be matched only with lower Paleozoic rocks now occurring above the Hinesburg thrust.

Thus the Hinesburg thrust (or an equivalent thrust occupying an even higher tectonic position!) must have been active in Middle Ordovician time (Zen, 1968, p. 138); the fault surface must have intersected the submarine depositional surface, so that clasts derived from the thrust plate were dumped into the contemporaneous sea. The relations suggest that the Hinesburg thrust was a gravity slide during Ordovician time. I suggest that at least during the Ordovician phase of its movement, the Hinesburg plate did not extend downdip into the Precambrian basement as reconstructed by Cady (1960); instead, the fault was very likely thin skinned (Fig. 5).

In northwestern Vermont the correlation of the gravity anomalies with surface geology is fair. The area of the Hinesburg plate, just west of the axial region of the Green Mountain anticlinorium, is an area of positive anomaly (Diment, 1968, Figs.30-3, 30-10). To the west, the area of imbricate thrust sheets and the adjacent foreland are areas of negative anomaly; a trough, exceeding -30 mgal in depth, comes south from Canada in the area northeast of St. Albans, approximately coincident with the axial region of the St. Albans synclinorium. Another trough, approaching -50 mgal in depth, is centered around Burlington, Vermont, and coincides with the Hinesburg synclinorium, which is the anticline south of St. Albans (Cady, 1945, p. 562). The cross-anticline coincides with the saddle in the gravity trough (Diment, 1968, Fig. 30-3).

Diment (1968, Fig. 30-10) showed a line of sharp, rudely circular positive anomalies extending east-southeast from northwest of Plattsburg, New York, to the area just south of St. Albans. He interpreted these bodies as a hidden group of dense intrusive rocks; the gravity saddle south of St. Albans, coincident with the occurrence of wildflysch-type conglomerates, may be caused at least in part by one of these hidden intrusive bodies.

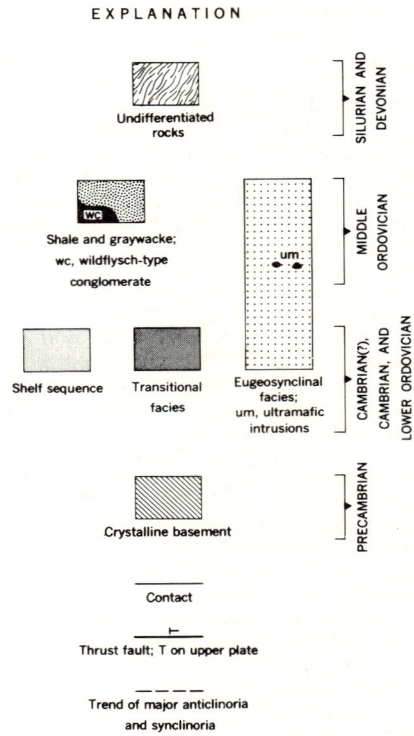

Figure 4. Generalized geologic map of north-central Vermont and adjacent New York, showing distribution of major lithofacies and tectonic elements discussed in the text. Compiled chiefly from Hawley (1957) and Doll and others (1961). See Figure 5 for sections A-A' and B-B'.

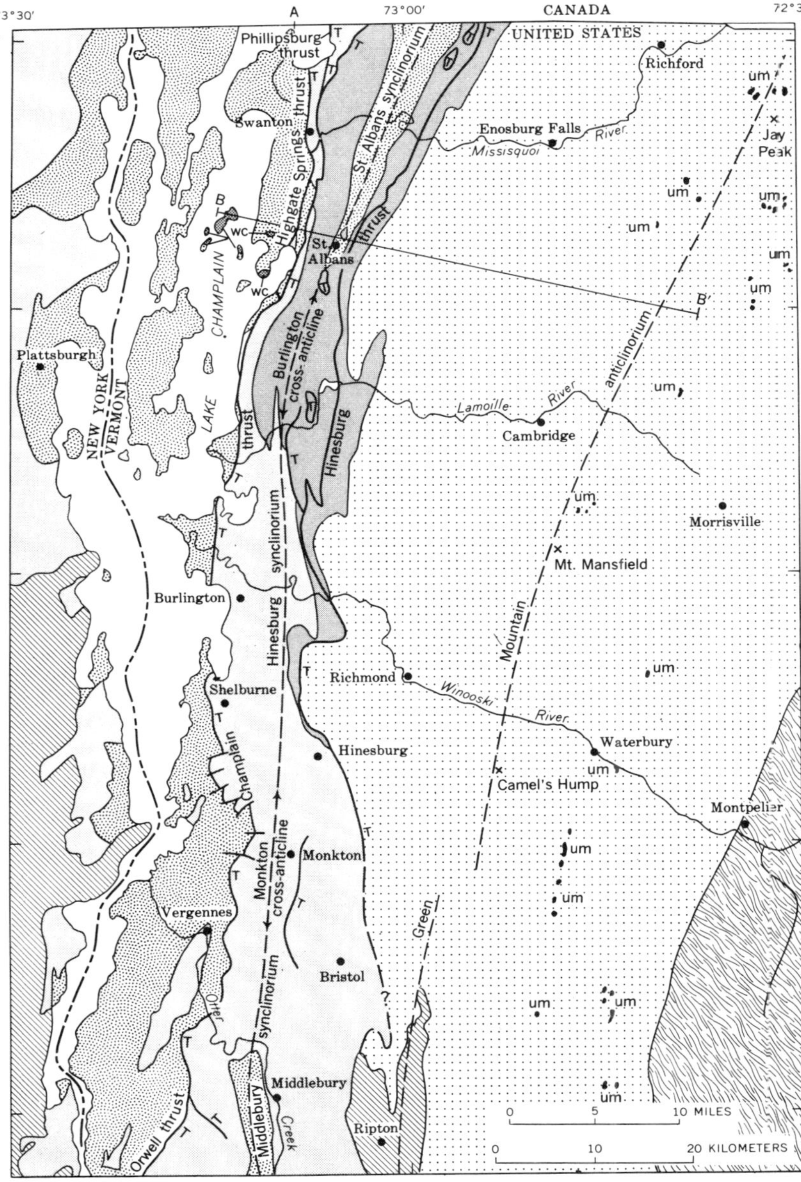

73°30'
A
73°00'
CANADA
72°30'
A'
45°00'
UNITED STATES
Phillipsburg thrust
Richford
Swanton
Highgate Springs thrust
St. Albans synclinorium
Enosburg Falls
Missisquoi River
Jay Peak
thrust
St. Albans
LAKE CHAMPLAIN
B
B'
wc
Plattsburgh
NEW YORK
VERMONT
Burlington cross-anticline
Hinesburg
thrust
Lamoille River
Cambridge
anticlinorium
Morrisville
Mt. Mansfield
44°30'
Burlington
Hinesburg synclinorium
Mountain
Richmond
Winooski River
Waterbury
Shelburne
Champlain
Hinesburg
Camel's Hump
Montpelier
Monkton cross-anticline
Monkton
Green
Vergennes
Bristol
Otter
synclinorium
Middlebury
Middlebury
Creek
Orwell thrust
Ripton
44°00'
um
0 5 10 MILES
0 10 20 KILOMETERS

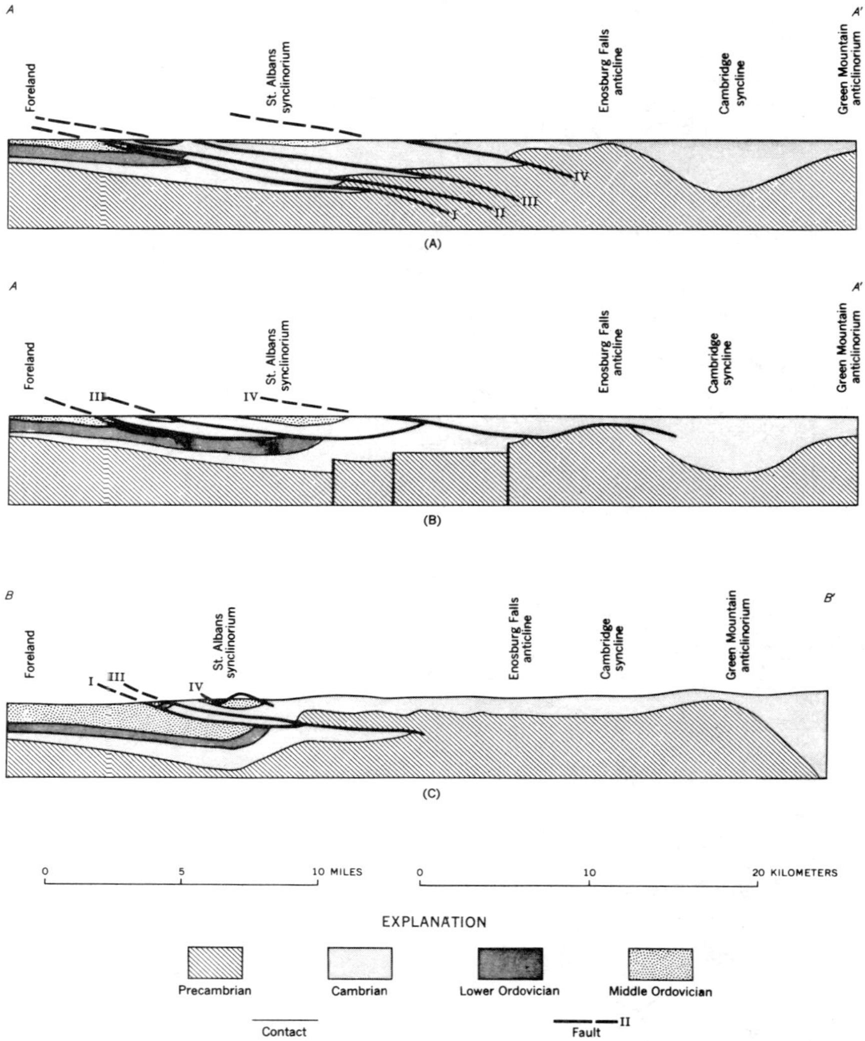

Figure 5. Three possible geologic cross sections across the western Appalachian orogen in the vicinity of the Vermont-Quebec border. Section A-A' lies along the border; section B-B' is about 25 km south of A-A' (see Fig. 4). (A), section A-A' after Cady (1960); (B), section A-A', present interpretation; (C), section B-B' after Doll and others (1961). In both (A) and (C), the thrusting involves basement rocks; in (B), the basement is not involved and "thin-skinned" tectonics are assumed to have been operative. No vertical exaggeration. Roman numerals designate individual thrusts.

Western Newfoundland

The Precambrian massifs of the northern Long Range and, to a much lesser extent, of Indian Head Range, dominate the Great Northern Peninsula of western Newfoundland (Fig. 6). Overlying the basement, and apparently dipping off it, is a shelf sequence of Cambrian and Ordovician sandstone and carbonate rocks (Schuchert and Dunbar, 1934; Williams, 1967b). The carbonate rocks pass upsection into calcareous shales and shaly limestones. These rocks are autochthonous; the youngest among these (the Clam Bank Formation on the Port-au-Port Peninsula) is Late Silurian and Devonian in age (Schuchert and Dunbar, 1934, p. 104; Rodgers, 1965).

In a region extending from Stephenville to Daniel's Harbour and also in the Hare Bay and Canada Bay region, basin deposits occur; fossils indicate that they are synchronous with rocks of the underlying and surrounding shelf sequence. Kay (1945) and later Rodgers and Neale (1963) proposed that the rocks are allochthonous, in structural relations analogous with those of the Taconic klippe. This fertile suggestion has been substantiated amply by recent fieldwork in these areas. Following Stevens (1968), the southern klippe will be called the Humber Arm klippe, and the one in the Hare Bay and Canada Bay area, the Hare Bay klippe. The geology of these areas has been summarized recently by Stevens (1970).

HARE BAY AREA. Rocks of the Hare Bay klippe consist of shale, graywacke, chert, greenstone, and ultramafic bodies; they are Lower Cambrian(?) to Ordovician (*see* Kay, 1967b) and resemble rocks of the Humber Arm klippe to the south. One unit, the Northwest Arm Formation, has yielded earliest Ordovician (Tremadoc) graptolites (Kay, 1967b, p. 3; Tuke, 1968).

Cooper (1937) and Betz (1939) first studied the rocks of the Hare Bay area; Cooper recognized that some of the mafic and ultramafic rocks are involved in low-angle thrusting. Tuke (1968) and Stevens (1968; 1970) mapped in the area; both authors agreed on the allochthonous nature of the rocks, on the existence of more than one geometrically independent slice, and on the location of the mafic and ultramafic rocks in the geometrically higher slices. Tuke (1968) also considered the higher slices to contain the older rocks, but Stevens (1969, written commun.; *see also* Stevens, 1970, p. 168-169) has recently discovered graptolites that indicate the ultramafic bodies to be among the younger rocks of the allochthon and to be post-Northwest Arm in age.

Note added in proof: See also the recent contribution on the Hare Bay klippe: Smyth, W. R., 1971, Stratigraphy and structure of part of the Hare Bay allochthon, Newfoundland: Geol. Assoc. Canada Proc., v. 24, p. 47-57.

The underlying autochthonous rocks range in age from Early Cambrian to Middle Ordovician (Schuchert and Dunbar, 1934, p. 17; Tuke, 1968, p. 505). These rocks are unconformably overlain by a Middle Ordovician mantling black shale (Goose Tickle Formation, Tuke, 1968, p. 506, = Englee Formation *of* Betz, 1939, p. 21). The autochthonous formations form a sedimentary sequence, and as a whole coincide in age with the allochthonous rocks. The mantling shale bears turbidite, belongs to a deeper water facies than the other units in this sequence, and contains boulders of rocks of the klippe in wildflysch-type conglomerates (Tuke, 1968). The emplacement of the allochthon, therefore, has been interpreted as the result of submarine gravity sliding, and the event is dated by the age of the black shale (Tuke, 1968). It is not certain,

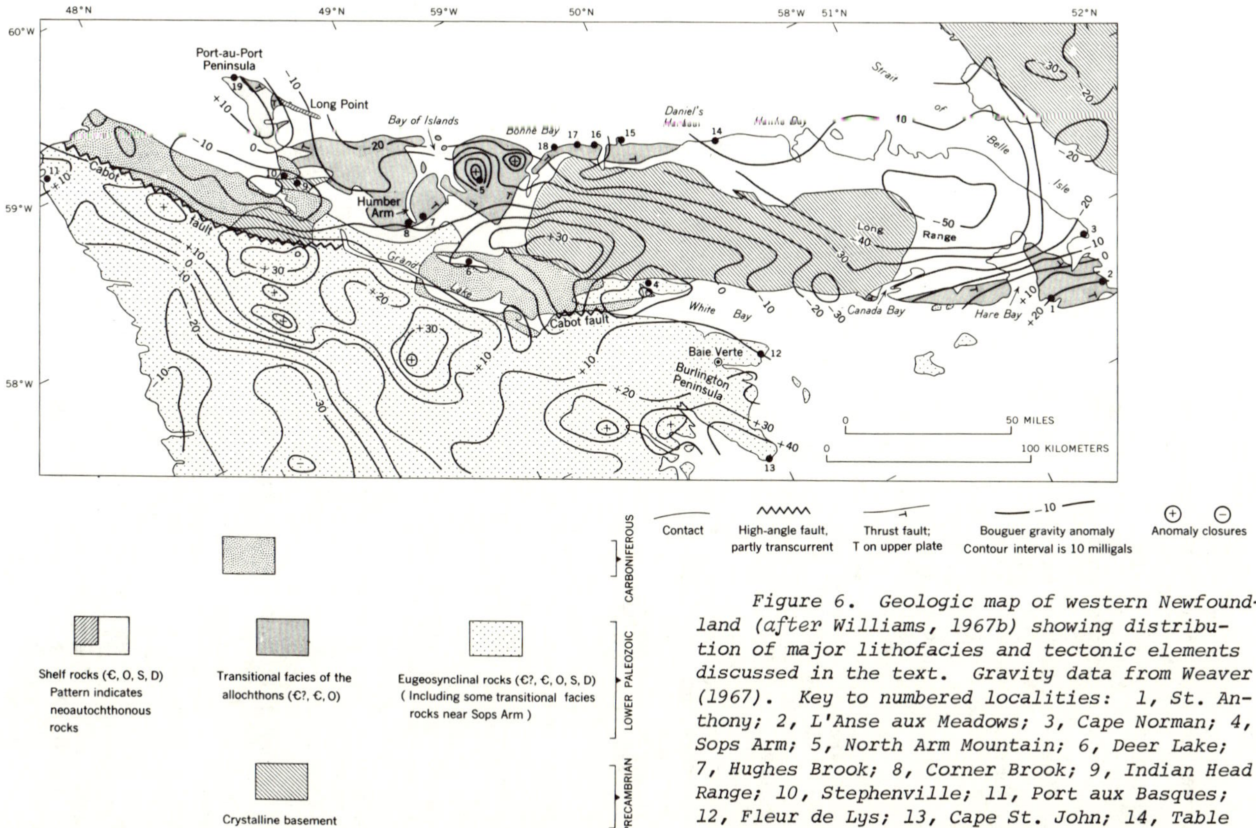

Figure 6. Geologic map of western Newfoundland (after Williams, 1967b) showing distribution of major lithofacies and tectonic elements discussed in the text. Gravity data from Weaver (1967). Key to numbered localities: 1, St. Anthony; 2, L'Anse aux Meadows; 3, Cape Norman; 4, Sops Arm; 5, North Arm Mountain; 6, Deer Lake; 7, Hughes Brook; 8, Corner Brook; 9, Indian Head Range; 10, Stephenville; 11, Port aux Basques; 12, Fleur de Lys; 13, Cape St. John; 14, Table Head; 15, Cow Head; 16, Broom Point; 17, Martin Point; 18, Green Point; 19, Cormorant Head.

however, that all the slices of the allochthon were emplaced by submarine sliding; the contact between the autochthon and the allochthon in the Canada Bay area (Betz, 1939, p. 21), for instance, might prove to be a hard-rock thrust fault similar to that suggested for the highest slice of the Taconic allochthon (Zen, 1967b).

HUMBER ARM AREA. The Humber Arm klippe occupies both shores of Humber Arm, and extends from Daniel's Harbour to the Port-au-Port Peninsula. Its geology is recently summarized by Stevens (1970). The klippe comprises two sedimentary facies. The bulk of the klippe, south of Bonne Bay, consists of at least four nested slices made up of shale, graywacke, volcanic rocks, and ultramafic and mafic intrusive rocks (Brückner, 1966; Cumming, 1967; Lilly, 1967; Stevens, 1970). These slices are mutually separated by zones of wildflysch-type conglomerates. Middle Cambrian to Early Ordovician fossils, definitely not reworked, were found in the thrust slices (Stevens, 1967). By lithic comparison with rocks of known age in the autochthon (Brückner, 1966, p. 146) and with a homologous sequence in the Taconic klippe (Stevens, 1970, p. 167), the Humber Arm klippe probably includes units as old as Early Cambrian and Cambrian(?).

The second sedimentary facies is found from Bonne Bay north to Daniel's Harbour. This is the facies of the Cow Head Group, including giant carbonate-clast conglomerates, bedded carbonate, and bedded shale and cherty shale. The rocks are well dated by fossils in the carbonate beds and carbonate-clast conglomerates; they are late Middle Cambrian to Early Ordovician (Kindle and Whittington, 1958, 1959, 1965). The rock type suggests an area of deposition different from that of the bulk of the allochthonous rocks in the Humber Arm area; unfortunately their mutual structural relations are still not known.

The autochthon consists of Lower Cambrian quartzite and Cambrian to Middle Ordovician carbonate overlain, possibly unconformably, by a lower Middle Ordovician (Whiterock) shale (Schuchert and Dunbar, 1934; Morris and Kay, 1966; Whittington, 1968). These rocks are coeval with rocks of the allochthon.

The contact zone between the autochthon and the allochthon includes a polymictic wildflysch-type conglomerate, which contains boulders derived from rocks of both the autochthon and allochthon (Brückner, 1966; Stevens, 1970). The emplacement of at least the lower, older slices of the klippe must have been submarine, presumably involving gravity sliding of soft sediments. The upper slices of the klippe contain large ultramafic bodies which presumably were solid during the emplacement of the slices; these younger slices may be conventional hard-rock thrust blocks.

As the klippe rests on lower Middle Ordovician rocks, its emplacement is no older than Middle Ordovician. In turn, in the Port-au-Port area, Rodgers (1965) showed that neoautochthonous rocks (the Long Point Formation) of Ashby-Porterfield age (Kay, 1969) rest with depositional contact on the klippe. If the Long Point Formation is entirely post-sliding in age, then the age of emplacement of the Humber Arm klippe is bracketed closely.

DISCUSSION. By comparing lithic sequences of early Paleozoic age, Rodgers and Neale (1963) concluded that rocks of the Humber Arm klippe could have been deposited in the area of Grand Lake and Deer Lake (Fig. 6), now underlain by metamorphosed shale, graywacke, and mafic volcanic rocks of Ordovician and older ages. These areas also include ultramafic rocks, so that the association seems right. Autochthonous lower Paleozoic rocks, definitely of the transitional facies, are also known to exist on the east side of the Great Northern Peninsula (*see*, for instance, Heyl, 1937; Betz, 1939).

Rodgers and Neale (1963) suggested, on the basis of lithostratigraphy and the enclosed ultramafic rocks, that either the Fleur de Lys Group, east of White Bay, or more likely the Baie Verte Group farther east, may be the undetached equivalent of the Hare Bay klippe. Tuke (1968) considered the probable depositional site for the Hare Bay klippe as north of the Burlington Peninsula, on strike with the present outcrop area of the Fleur de Lys Group and its associated ultramafic bodies.

It is clear, in general, that possible depositional sites for the allochthonous rocks ought to have become structural highs, and that it is unlikely for such sites to be now occupied by correlative rocks. Also, if the Cabot fault zone should prove to be a major strike-slip fault, as has been suggested, for example, by Belt (1968), or by Williams and others (1970), then the depositional sites for the rocks of both klippen of Newfoundland should not be sought for in the projected updip direction across the fault.

High-angle faults are known on the Great Northern Peninsula in Newfoundland. Cumming (1967) showed that west of the Precambrian massif of the northern Long Range the prevailing dip-slip sense is west side down, but opposite sense of dip slip exists at Cape Norman at the northwestern extremity of the peninsula (Cumming, 1967, Fig. 1; Tuke, 1968, p. 504). Williams (1967b) showed these as well as numerous other faults on the Great Northern Peninsula, both in the klippen and in the autochthon; however, some of the faults may be low angle (Johnson, 1941, p. 144).

Tuke and Baird (1967, p. 9), Brückner (1966, p. 150), and McKillop (1963, p. 89) suggested that the high-angle faults in western Newfoundland followed the emplacement of the klippen. McKillop (1963, p. 81) pointed out, however, that the high-angle faults do not enter the allochthonous terrane, and Cooper's (1937, p. 12) schematic cross section shows high-angle faults truncated by his Hare Bay thrust, which is the base of one of the gravity slices. Some of the high-angle faults thus probably had at least an Ordovician beginning.

The gravity field of western Newfoundland (Figs. 3, 6) has been studied recently by Weaver (1967); the results correlate well with bedrock relationships. To summarize: the area of the Humber Arm klippe is on the southeast flank of a deep and extensive trough of Bouguer anomaly. The deepest anomaly, more than -50 mgal, in the northwestern part of the peninsula, is in an area of extensive shelf deposits. The Hare Bay region is on a steep gravity gradient between the deep trough and an offshore high to the southeast. Two ridges of positive anomaly extend south-southwest from the area of Burlington Peninsula. The eastern ridge is better defined and higher; it extends across the southern Gulf of St. Lawrence to become the major gravity ridge of the Maritime provinces (Fig. 3; *see* Goodacre and others, 1969). The less sharply defined western ridge extends from the Hare Bay area through the White Bay area to the Port-au-Port Peninsula and connects with the gravity ridge of Gaspé Peninsula through a low gravity ridge in the Gulf of St. Lawrence. As a whole, this gravity ridge parallels the regional tectonic trend, and broadly coincides with the zone from which Rodgers and Neale (1963) would derive the Humber Arm klippe. The northward extension of this gravity ridge places it where Tuke (1968) proposed a tectonic source for the Hare Bay klippe.

Weaver (1967, p. 244-246) computed that, to agree with the gravity data, the ultramafic sheets in the Humber Arm klippe are about 1 km

thick or less, consistent with the supposition that the bodies are part of rootless gravity slides. The exception, the North Arm Mountain intrusive body, northeast of the Bay of Islands, has a local anomaly of +40 mgal, requiring an ultramafic and mafic body extending to about 4.6 km (2.8 mi) below the surface. The thickness is not necessarily too great for the allochthon, and, in fact, agrees with the projected lopolithic structure of the ultramafic sheet (H. D. Lilly, quoted *by* R. K. Stevens, 1968, oral commun.).

The gravity gradient between the regional trough and the ridge is smooth and moderately steep. The width of the gradient zone suggests that the source of the anomaly is in the lower part of the crust, perhaps 20 km below the surface.

Taconic Klippe and Environs

The Taconic klippe consists of rocks of the shale-graywacke association and contains only minor mafic volcanic rocks (Fig. 7). The klippe consists of six or possibly more separate, but nested, slices; the younger slices now occupy the higher geometric positions. The lower slices, containing the younger stratigraphic units, show abundant evidence of soft-rock deformation. These slices rest on autochthonous carbonate, quartzite (shelf sequence), and mantling black shale. The mantling black shale includes graywacke and a polymictic wildflysch-type conglomerate many of whose boulders, pebbles, and chips are lithically identifiable as part of the allochthonous sequence (*see* Bird, 1969, for further discussion).

Some of the allochthonous rocks are fossiliferous, ranging in age from Early Cambrian to Middle Ordovician (Wilderness Stage, *see* Bird and Rasetti, 1968; summary and references *in* Zen, 1967a). As the autochthon has yielded Cambrian and Early and Middle Ordovician fossils, the two sequences are completely synchronous.

The age of emplacement of the Taconic klippe is determined by the presence of graptolites in the matrix of the autochthonous wildflysch-type conglomerate, correlated by Berry (1970, p. 310) with the Canajoharie fauna of Riva (1969a, b), and by relating the derivation of the conglomerate with the emplacement of the klippe (Zen, 1967a, p. 68). This age is slightly younger than the age of the youngest fossils in the allochthon. The upper age limit of the emplacement of the geometrically lower slices is fixed by the Upper Silurian or basal Devonian rocks of Becraft Mountain near Hudson, New York, which lie astride the contact between the allochthon and autochthon. The upper slices were probably emplaced in Late Ordovician or Early Silurian time (Zen, 1969).

The possible site of deposition for rocks of the Taconic klippe has been examined (Zen, 1961, p. 330; 1967a, p. 56); a likely site is over the present Green Mountain and Berkshire massifs, in the intervening areas, and possibly over the Housatonic massif.

High-angle, longitudinal faults that have both normal and reverse senses characterize the areas of the autochthonous shelf rocks both east and west of the allochthon. Zen (1968) reviewed the data; much additional information is given by Thompson (1967). One effect of these high-angle faults was to contribute to the subsidence of the shelf; the Precambrian massifs of the Green Mountain and Berkshire Hills may owe their structural relief at least in part to faulting of the same sort (Zen, 1967a, p. 71). These faults are likely primarily dip-slip faults.

The amount of throw cannot be determined with precision, but estimates are possible. Thompson (1967) recorded the juxtaposition of Precambrian basement rocks and Lower Ordovician carbonate rocks across a high-angle fault in western Vermont, and both fault blocks are mantled over by the Middle Ordovician black shale. The stratigraphic throw must be at least 500 to 800 m (about 2,000 ft), the thickness of the Cambrian section of the shelf sequence; the minimum cumulative uplift is the entire thickness of the pre-black shale part of the shelf sequence, or about 1.5 km (5,000 ft). The total net vertical movement of the fault blocks may well be at least 3 to 5 km (2 to 3 mi).

The field relations of these faults are compatible with active movements in Middle Ordovician time (Zen, 1967a, 1968). Some of the faults may have started near the end of the Early Ordovician, but the episode

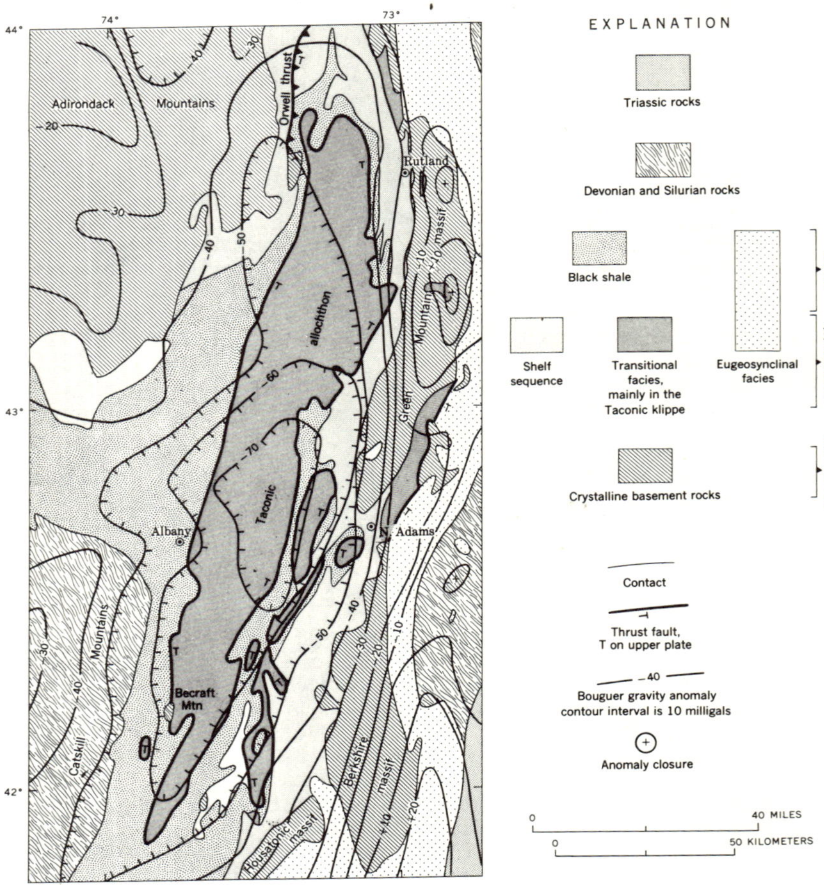

Figure 7. Geologic map of the region of the Taconic klippe in western New England and eastern New York, showing distribution of major lithofacies and tectonic elements discussed in the text. Geology modified from Zen (1967a); gravity data from Woollard and Joesting (1964) and Diment (1968).

of movement lasted at least into late Trenton time, about the time of emplacement of the klippe. Thompson (1967, p. 88) suggested possible Middle Ordovician volcanic activity along the faults in western Vermont.

The Taconic klippe coincides with a large Bouguer gravity trough delineated by Diment (1968); the -70 mgal closure nearly coincides with the trace of the highest slice of the allochthon (Fig. 7; Zen, 1967a, p. 72). To the east, the Green Mountain massif coincides with a sharp gravity ridge locally exceeding +10 mgal. The half-width of the gradient zone is about 15 km, suggesting a source of anomaly within the crust. The coincidence between the Berkshire massif and the gravity high is good at the north end, but deteriorates to the south, where the gravity ridge is displaced southeastward to the east flank of the anticlinorium. This displacement has been correlated (Zen, 1967a, p. 72) with the large westward transport of the southern part of the massif itself.

Northern New Jersey and Eastern Pennsylvania

A wide belt of Middle to Upper Ordovician dark shale, sandstone, and graywacke, the Martinsburg Formation (Gray and others, 1960), crosses the southeastern part of Pennsylvania in a west-east direction parallel to the Appalachian trend and north of the Triassic basin (Fig. 8). These rocks are the mantling black shale (and basal limestone) over the carbonate-quartzite units of the shelf sequence that range in age from Early Cambrian (Hardyston Quartzite) to Middle Ordovician (Jacksonburg Limestone, Chambersburg Limestone). Within the area of "normal" Martinsburg, there occur patches of red and green shale, graywacke, carbonate-clast conglomerate, and locally even pillow lava (Stose, 1946, p. 673; Bricker, 1960). These rocks are of anomalous sedimentary facies and locally have yielded Early Ordovician graptolites; because they now overlie or are enclosed within younger rocks of the Martinsburg Formation, they must be allochthonous (Willard, 1943, p. 1097-1101; Stose, 1946). Kay (1941) suggested that a Taconic-type klippe exists here. On the Geologic Map of Pennsylvania (Gray and others, 1960), no klippe is shown; rather, the volcanic rock, conglomerate, and red shale which typically yielded Early Ordovician graptolites are shown as isolated patches of small dimensions, confined to the area immediately north of Lebanon, and of unspecified structural relations. The allochthonous rocks may consist of erosional remnants of a once large klippe, or pieces that slid into place separately, perhaps as giant blocks in a wildflysch-type conglomerate, during the Middle Ordovician as part of a submarine sliding episode (MacLachlan and Root, 1966; Carswell and others, 1968; Alterman, 1969; Platt and others, 1969).

Three outliers of anomalous Ordovician shale occur in central New Jersey between crystalline Precambrian rocks to the north and Triassic rocks to the south. The normal rocks bear Normanskill fossils and consist of gray silty shale and second-cycle graywacke. The anomalous rocks here are purple slate, carbonate-clast polymictic as well as monomictic conglomerate interbedded with green and black banded slate, interbedded green chert and slate, and red slate containing beds of thin green quartzite. These rock types have been described by Stose (1946, p. 689) and Drake (1969, p. 95); they match Cambrian and Lower Ordovician rocks in the Taconic klippe as well as the anomalous rocks of southeastern Pennsylvania. Field relations suggest that all the anomalous rocks could be blocks in a gigantic field of wildflysch-type conglomerate having the Normanskill shale and graywacke as the matrix.

Ordovician fossils have been known in the Jutland outlier since the

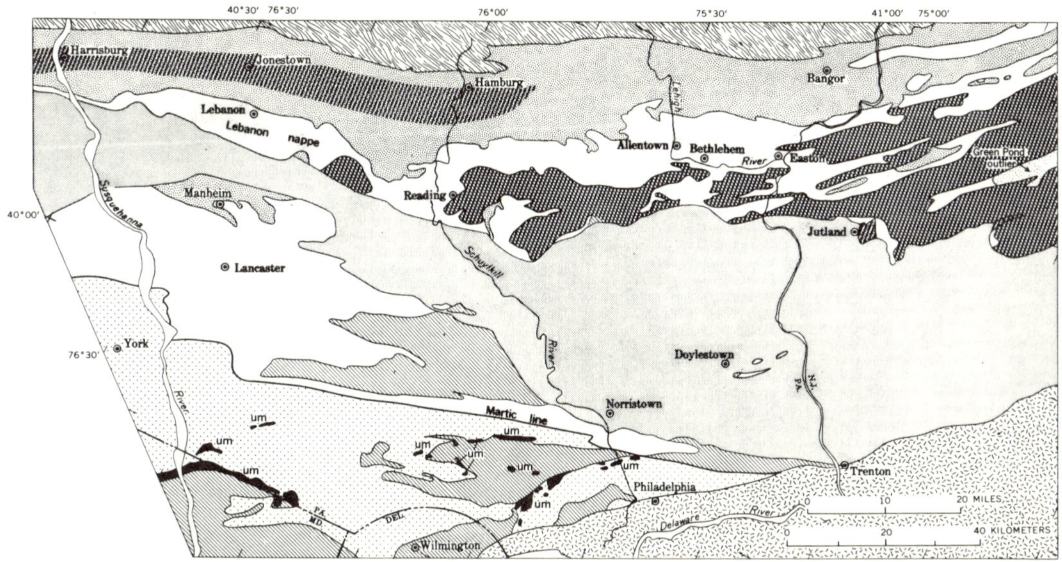

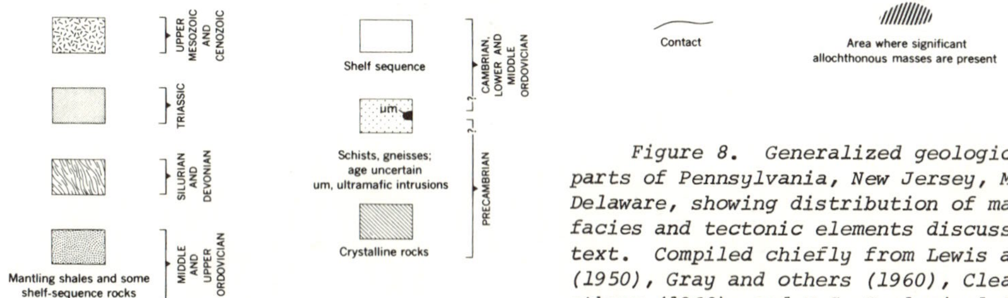

Figure 8. Generalized geologic map of parts of Pennsylvania, New Jersey, Maryland, and Delaware, showing distribution of major litho-facies and tectonic elements discussed in the text. Compiled chiefly from Lewis and others (1950), Gray and others (1960), Cleaves and others (1968), and U.S. Geological Survey (1968).

report of Bayley and others (1914, p. 12). These fossils are mainly Normanskill graptolites, but older Deepkill graptolites have been confirmed recently by Neuman and Dodge (*in* U.S. Geol. Survey, 1964, p. A-83). Because the Jutland outlier lies structurally above the carbonate units of the Kittatinny Limestone (Lower Ordovician), and possibly also above the Jacksonburg Limestone (Middle Ordovician)(Bayley and others, 1914; Barnett, 1965), the Deepkill graptolites are clearly not in a normal sequence, and the Normanskill also may not be.

Drake (1969) showed that the Precambrian rocks immediately north of the Jutland and related outliers are themselves part of a large far-traveled klippe. This allochthonous relation extends at least from north-central New Jersey southwest to the termination of the Reading Prong in Pennsylvania. The movement of this klippe of Precambrian rocks cannot be dated closely, but it is clearly post-Martinsburg, presumably younger than the wildflysch-type conglomerates at Jutland (the formation of the conglomerates is probably early Martinsburg). The thick section of Middle to Upper Ordovician Martinsburg graywacke and shale in north-central New Jersey might record the initial development of the Precambrian klippe as well as the earlier movements that led to the slide masses of Paleozoic sedimentary rocks (Drake, 1969, p. 92).

The original depositional sites for the allochthonous lower Paleozoic masses of Pennsylvania and New Jersey are unknown. Kay (1941) early suggested that the Martic Line might be the fault that produced the thrust mass, so that the depositional sites would be in the area now underlain by the Glenarm Series of the Maryland Piedmont. Drake (1969, p. 94) suggested that the Cocalico Shale of southeastern Pennsylvania, immediately south of the Triassic basin, but north of the Martic Line, could be the "source of the klippen"; however, the true structural and stratigraphic relations of this shale unit to the regional geology are unclear. The Cocalico Shale is a lithically complex unit and includes gray, purple, and green shale, vitreous quartzite, and volcanic tuff; perhaps it is not a simple sedimentary sequence, but includes exotic rocks (*see* Stose, 1946, p. 690; Jonas and Stose, 1930, p. 39).

The depositional sites of the allochthonous masses must have been more seaward, that is, southeast of the junction of the lower Paleozoic carbonate-shale facies, which in this area has been identified with the Kinzers and Conestoga Formations near Lancaster, Pennsylvania (Rodgers, 1968; Kauffman and Campbell, 1969). This area is considerably southeast of the main outcrop belt of the Cocalico Shale near Manheim, Pennsylvania (Fig. 8). Although a depositional site in the Cocalico cannot be ruled out, the Cocalico itself may be allochthonous (Stose, 1946, p. 690-692) or a large wildflysch-type conglomerate field, so it is worthwhile to consider possible depositional sites over rocks now exposed south of the Martic Line or on now-eroded terrane once carried by a thrust plate to the northwest.

Rocks southeast of the Martic Line belong to the Glenarm Series, chiefly the Wissahickon Formation, of uncertain age. Hopson (1964) concluded that the Wissahickon is no younger than Early Cambrian, much older than most of the allochthonous masses. The Glenarm is approximately of the same sedimentary facies as the allochthonous masses of Pennsylvania and New Jersey, and rocks of the Glenarm Series, or rocks that once overlay it, could have furnished the exotic blocks now embedded in the Martinsburg Formation. This would be true regardless of the structural interpretation of the Martic Line itself (Wise, 1970); for instance, a simple hinged uplift along this line in Middle Ordovician time could have triggered gravity sliding of the cover rocks.

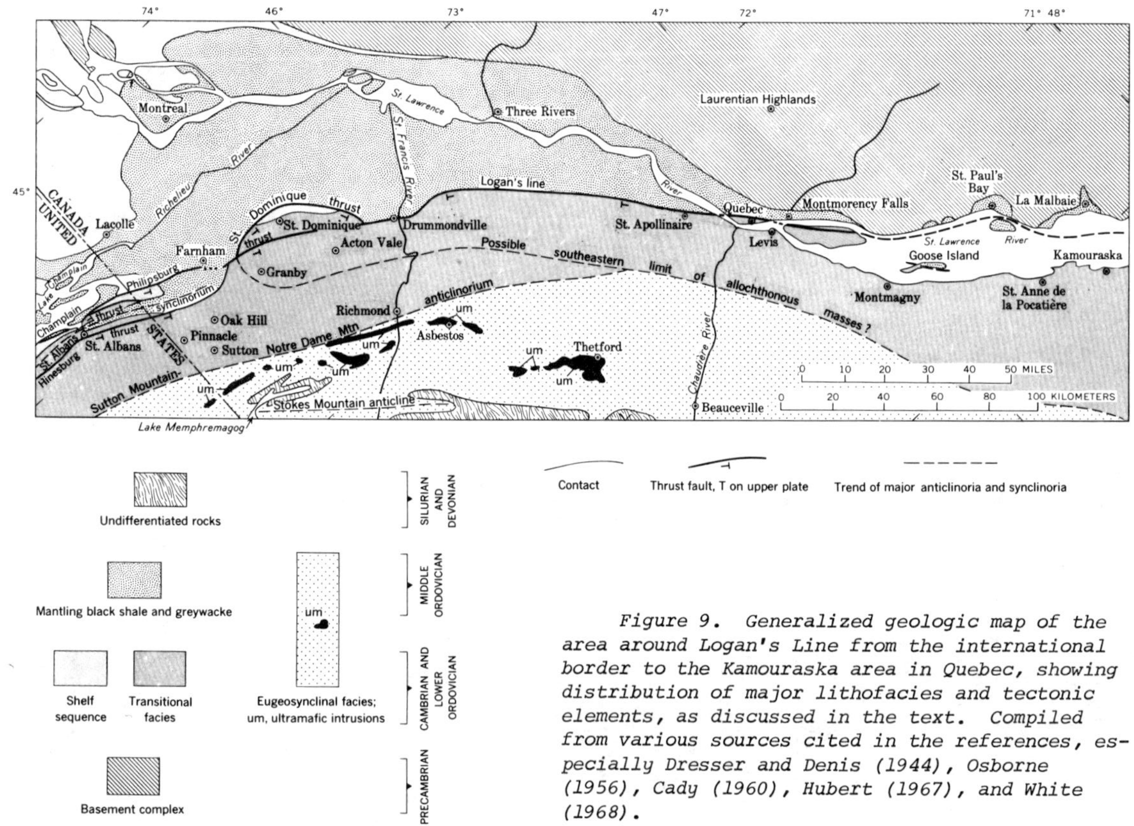

Figure 9. Generalized geologic map of the area around Logan's Line from the international border to the Kamouraska area in Quebec, showing distribution of major lithofacies and tectonic elements, as discussed in the text. Compiled from various sources cited in the references, especially Dresser and Denis (1944), Osborne (1956), Cady (1960), Hubert (1967), and White (1968).

On the other hand, M. W. Higgins (*in* U.S. Geol. Survey, 1969, p. A29) found that the field and laboratory evidence used to assign a late Precambrian age to the Glenarm Series is not compelling. Many supposed intrusive rocks, whose zircon ages were crucial to Hopson's (1964) synthesis, are regarded by Higgins and independently by Crowley (1969) as metamorphosed volcanics, so the zircon ages are maximum, rather than minimum, ages. If so, then the Glenarm could include rocks as late as Ordovician; furthermore, the facies then would correspond to the lower Paleozoic volcanics-rich eugeosynclinal sequence in the northern Taconides. If the allochthonous masses in the Martinsburg Formation were never much more extensive than they are today, these rocks could have been derived from a lower Paleozoic Glenarm Series.

Drake (1969, p. 100-131) reviewed the regional gravity and aeromagnetic relations in north-central New Jersey and adjacent Pennsylvania, and explained these in terms of the bedrock geology. He pointed out that the allochthonous Precambrian mass is within a large area of Bouguer gravity low (-60 mgal); this same gravity trough also includes the area of allochthonous Paleozoic masses of southeastern Pennsylvania, whereas those of New Jersey are on the slope between the trough and a flanking gravity ridge (+40 mgal) to the southeast. The gravity trough, in fact, is southwest of and continuous with that of the Taconic area, and the gravity ridge is the continuation of the ridge extending south from the Green Mountain massif and the east flank of the Berkshire massif (Fig. 3). The gravity ridge is centered south of the Triassic rocks in an area now covered by the Coastal Plain sediments. Bromery (*in* U.S. Geol. Survey, 1966, p. A-70) concluded from the geophysical data that the Precambrian rocks in the gravity trough are thin and thus presumably unrooted.

Relation of the Allochthonous Rocks along Logan's Line

LOGAN'S LINE. Although Logan's Line is defined primarily as a structural break (Logan, 1863, p. 233-234; Clark, 1951; Neale and others, 1961), it is recognized, in part, by the sharp stratigraphic contrast across it (Fig. 9). On the north and west side of the line, the rocks are autochthonous Precambrian basement rocks, lower Paleozoic carbonate rocks, and Middle Ordovician carbonate rocks and mantling black shale. On the south and east side in Quebec, the rocks are red, green, and black shale, graywacke, sandstone, mafic volcanics, and carbonate-clast conglomerate.

Logan's original definition of the line extended it from near Quebec City southwestward to the Vermont line, and eastward under the St. Lawrence River, to re-emerge to "within about eighty miles of the extremity of Gaspé, where it again comes upon land," presumably along the Lower Ordovician—Middle Ordovician contact mapped by Enos (1969) and by Riva (1968) at Marsoui. The entire northwest front of the Taconides (Logan's Line for the most part) in Quebec is likely a zone of horizontal transport, and the rocks to the southeast are allochthonous. These rocks can be conveniently described under areal subheadings.

KAMOURASKA AREA. Arkose, graywacke, and shale occur about 50 km northeast of Quebec City, along the south shore of St. Lawrence River. These rocks are Early Cambrian to Early Ordovician (Rasetti, 1946), part of the "Sillery" formation; their field relations have recently been studied by Hubert (1965, 1967). The rocks occur in three nested thrust slices; the older slices, occurring to the northwest, contain the younger rocks. These younger, Ordovician rocks also include carbonate-clast conglomerates, similar in occurrence to the Levis conglomerate in the Quebec City area (*see* Rasetti, 1946, p. 702). Hubert (1965, p. 87) showed that for the one studied formation, the source direction of the carbonate-clast conglomerate was northwest (craton), whereas that of the graywacke and shale beds was south and east (basin; *see* additional information *in* Hubert and others, 1970).

Lower Cambrian rocks of the Kamouraska area are in contact with Middle Ordovician rocks on Goose Island (Hubert, 1967, p. 36). Thus, although the tectonic position of these Middle Ordovician rocks has not been stated, the emplacement of the slices of the Kamouraska area is almost surely a Middle Ordovician event by analogy with relations in the Quebec City area. The rocks of the Kamouraska area are contiguous to the southeast, across structural trend, with probably autochthonous eugeosynclinal rocks of the same general age; however, the structural relations are not clear from published data, and at least the upper slices might not be totally detached.

QUEBEC CITY AREA. The general relations have been summarized by Osborne (1956). The northern part of the area is underlain by the Precambrian rocks of the Laurentian Highlands, part of the craton. Although lower Paleozoic rocks of the shelf sequence are absent in outcrop, their former presence is shown by the abundant carbonate clasts, ranging in age from Early Cambrian to Early Ordovician, in the allochthonous shale and graywacke sequence southeast of Logan's Line. Middle Ordovician black shales and associated limestones that mantle the basement also extend into the foreland area to the southwest, where shelf rocks, deposited on a craton, are known.

Southeast of Logan's Line, and extending southwest on trend to the vicinity of Drummondville, the "Sillery Group" occurs; it contains Lower Cambrian to Lower Ordovician rocks (Rasetti, 1946; Osborne, 1956; Osborne and Riva, 1966). These rocks are dominantly shale and graywacke, but include carbonate rocks — both bedded and conglomeratic — which are especially abundant in the Upper Cambrian and Lower Ordovician formations (Lauzon and Levis). Volcanic rocks, including pillow lava, are locally found (Osborne and Riva, 1966). Immediately south of Quebec City, at least, the "Sillery" rocks are in several detached thrust slices (klippen) showing southeast-to-northwest movement sense (St-Julien, 1968); all the Taconide zone rocks south of Logan's Line in this region are allochthonous.

The lowest and oldest thrust slice in the Quebec City area consists of Normanskill-age black shale, the Quebec City Formation (St-Julien, 1968, p. 1351, also Fig. 2). One of the several chronostratigraphically distinct polymictic wildflysch-type conglomerates of this formation, the Citadel Formation (Osborne, 1956), was dated as Trenton, or possibly younger, by Stäuble (1952, p. 315), but was dated recently as Normanskill (St-Julien, 1968, p. 1351). This rock was interpreted by Bailey and others (1928) as made up of submarine slide deposits; however, this unit now seems explained best as consisting of deposits in front of one of the gravity slides of the "Sillery Group" (St-Julien, 1968).

High-angle faults are common in the St. Lawrence Lowland near Quebec City; perhaps the best known of these is the fault at Montmorency Falls, which juxtaposes Precambrian and Middle Ordovician rocks (Emmons, 1841; Logan, 1863, p. 234). Osborne (1956, p. 166) summarized data on the faults, and concluded (1956, p. 169) that their ages cannot be determined. An age roughly synchronous with the gravity sliding is not incompatible with the stratigraphic relations. High-angle faults, apparently of Middle Ordovician age and having south-side-down dip-slip sense, have been deduced from subsurface data on Anticosti Island (Roliff, 1968, p. 33).

GRANBY-ACTON VALE AREA. In the area around Granby, Quebec, graywacke, shale, quartzite, and sparse limestone of the "Sillery" formation occur, surrounded by Middle Ordovician black shale that overlies pre-Middle Ordovician Paleozoic carbonate rocks of the shelf (*see* Clark, 1964a, b). The "Sillery" formation here constitutes a lithostratigraphic sequence that closely resembles the sequence of the Taconic klippe and that of the Quebec City-Drummondville area. However, these "Sillery" rocks are of unknown age, and their structural relations to the surroundings are likewise unclear. Clark and Eakins (1968) suggested that they are part of a submarine gravity-driven allochthon.

Cady (1960, p. 558) raised the problem that the clastic "Sillery" rocks, having a northwesterly cratonal source (Osborne, 1956, p. 197), probably could not have traversed a shelf. However, Osborne's reference is only to the Charny Formation (Lower Cambrian), which is older than the oldest unit of the shelf in this area (Cady, 1960, Pl. 3). The problem parallels that for the Taconic allochthon whose oldest unit, the Rensselaer Graywacke, has a westerly source; however, the Rensselaer, which is lithically similar to the Charny, is of a lithic facies distinct from the overlying, partly fossil-bearing Cambrian and Ordovician strata. This problem has been discussed elsewhere (Zen, 1967a, p. 63-64). Unless it can be demonstrated that the post-Charny rocks of the "Sillery" formation of the Granby area also had a northwesterly source, no real problem exists.

Clark (1964b) reported badly contorted Middle Ordovician black shale, called the St. Germain Complex, on the west side of the Granby area. These beds are locally the wildflysch type. Clark and Eakins (1968) interpreted the contorted nature of these shale beds as being the result of kneading action produced by the sliding of a Granby klippe of the basin facies over the soft sediments. The St. Germain Complex here is early Middle Ordovician (Clark and Eakins, 1968; *compare* Riva, 1966, 1969b), which, therefore, dates the event of sliding.

DEPOSITIONAL SITES FOR THE ALLOCHTHONOUS ROCKS. The area of original deposition of the transported rocks along Logan's Line cannot be stated yet. These rocks must have moved in from the south and east, possibly from the area along the Sutton Mountain—Notre Dame Mountain—Shickshock Range anticlinorial region, where rocks of the same general lithofacies and same age are exposed and where complex history of deformation is known (*see*, for instance, Osberg, 1965; Sikander and Fyson, 1969). Many problems must be solved before a definite statement on the depositional sites can be offered. For instance, St-Julien (1968) mapped an extensive belt of wildflysch-type conglomerate that runs northeast from the north end of Lake Memphremagog to about 50 km (30 mi) south of Quebec City. This belt is southeast of the Sutton Mountain anticlinorial axis and separates rocks of the Caldwell Group (Cambrian-Ordovician) to the northwest from younger rocks to the southeast. This belt of rocks surely must play an important, although still obscure, part in the history of the transported rocks along Logan's Line. If the belt defines, as seems possible, the southeast or trailing edge of an extensive composite klippe, then the depositional site must be much farther southeast, and the distance of transport must be at least 40 km (30 mi).

R. K. Stevens (1969, written commun.) pointed out that a serious room problem exists on Gaspé Peninsula unless the ultramafic rocks (Mt. Albert mass and others) are not rooted locally; he suggested that the depositional site might be found on the south shore of the peninsula. Lack of a recognizable trailing edge for the transported rocks (largely hidden under younger rocks of the Connecticut Valley–Gaspé synclinorium) means that no more than a few kilometers of transport, corresponding to the depth of re-entrant of the autochthon-allochthon contact along the Gaspé north shore (Riva, 1968) can be proven. Solution of the structural role of the wildflysch-type conglomerate band (St-Julien, 1968) south of Quebec City will have important bearing on the problem on Gaspé Peninsula.

Another, indirect indication of the scope of transport is the absence of a pre-Middle Ordovician shelf sequence along Logan's Line in Quebec, although rocks obviously belonging to this sequence are abundantly preserved in carbonate-clast conglomerates of a northerly derivation and in the subsurface of Anticosti Island (*see* Append. 2). A shelf of uncertain width must have formed and, subsequently, become covered by later sedimentation and sliding, or possibly in part destroyed by tectonic loss of crust. To the extent that large horizontal transport seems certain in the area, the original depositional site for the rocks probably should be sought in the area at least as far southeast as the Sutton Mountain–Notre Dame Mountain anticlinorial region; the site, in fact, could be hidden under the Silurian and Devonian rocks of the Connecticut Valley–Gaspé synclinorium.

GRAVITY DATA. The general disposition of belts of Bouguer anomalies found in Newfoundland by Weaver (1967) is also seen in Quebec south of the St. Lawrence River, and in adjacent areas of New Brunswick (Fig. 3; Innes and Argun-Weston, 1967). A gravity ridge is just west of, or coincides with, the axis of the Sutton Mountain–Notre Dame Mountain anticlinorium, underlain by Cambrian and Ordovician metamorphic rocks, from the Canada-United States border northeast; this broad coincidence is observed as far north as Lake Temiscouata (Figs. 1, 3). A zone where detailed gravity relations are uncertain then intervenes; on the Gaspé Peninsula the gravity high again coincides with the topographic high along the Shickshock Range, and, following closely the trace of the post-Taconic unconformity on the north flank of the Connecticut Valley–Gaspé synclinorium (McGerrigle and Skidmore, 1968), reaches the Gulf of St. Lawrence at Cape Gaspé.

North of the gravity ridge and closely following the St. Lawrence River is a series of gravity troughs (Fig. 3). The most extensive of these troughs, in the area between Matane and Kamouraska, has a minimum value less than -50 mgal. This trough includes much of the allochthonous rocks of the Quebec City area as well as all of those in the Kamouraska area as mapped by Hubert (1965). Another deep trough along the same trend occurs to the northeast in the Matane area, immediately northwest of the gravity peak of the Shickshock Range.

Thompson and Garland (1957, p. 127) explained the trough near Quebec City as caused by excessively thick sedimentary rocks, in part resulting from thrust sheets. Innes and Argun-Weston (1967, p. 78) pointed out that there is no general correlation between thick Paleozoic sections and gravity lows (nearby areas of Precambrian shield show even greater negative anomalies), and the density contrast among exposed bedrock types is not enough to account for the anomaly. Innes and Argun-Weston (1967, p. 78) inferred that the gravity contrast reflects fundamental lithic differences between crustal columns in the respective areas.

The gravity ridges and troughs are not well defined in the Granby–Acton Vale area. A large gravity ridge coincides with the west limb of the Sutton Mountain anticlinorium (Dominion Observatory Ottawa, 1967), and vaguely defined gravity troughs and saddles occur west of the ridge, connecting the deep troughs along the St. Lawrence River with those in western New England. The area of basin rocks at Granby approximately coincides with one of the broad troughs (Innes and Argun-Weston, 1967, p. 73), but the relations are complicated by the presence of east-southeast-trending intrusive bodies of the Monteregian Hills series, whose sharp, localized positive anomalies (Diment, 1968) presumably distort the shapes, locations, and magnitudes of the regional Bouguer anomalies.

Interpretation

Major Paleogeographic Elements

The major paleogeographic elements that affected the evolution in the Taconide zone have been mentioned. To summarize: the craton was located on the northwest side, skirted by a wide area of shallow-water, supratidal, intertidal to subtidal carbonate deposition in an environment of platform and shelf. This shelf environment extended with little change at least from northwestern Newfoundland to Alabama, and persisted for long periods of time. The shelf facies passed into a eugeosynclinal facies across a narrow transitional belt. The transitional sediments were derived both from the shelf direction and from the opposite, presumably oceanward direction (Hubert, 1965; Zen, 1967a). This and the presence of chert, turbidites, radiolaria (Ruedemann and Wilson, 1936), and graptolites indicate a deep-water environment. I suggested (1967a) that this belt corresponded with the axial region of the depositional trough. Farther seaward of this belt was the main eugeosynclinal zone, containing islands, both volcanic and nonvolcanic (*see* Kay, 1967a; Pavlides and others, 1968; Berry, 1968; Thompson and others, 1968; Naylor, 1968). Sedimentary rocks in this zone are marked by rapid facies change, by an abundance of both subaerial and submarine volcanic material, by the near absence of carbonate rocks, and by enormous thicknesses. There is no sharp boundary between the eugeosynclinal and the transitional facies, but the contrast between the latter and the shelf facies is sharp, is commonly marked by a zone of carbonate-clast conglomerates, and was spatially stable over long periods from the earliest Cambrian to the end of the Early Ordovician or well into the Middle Ordovician.

The age of the rocks in the craton requires comment. Rodgers (1968, p. 148) pointed out that the southeastern limit of radiometrically determined 1 (±) b.y. Precambrian crystalline rocks on the east coast of North America is sharply defined. This limit very nearly follows the line of Precambrian massifs that we mentioned previously. In Newfoundland the limit is the northern Long Range and Indian Head Range (Brückner, 1966, p. 143). Southeast of a line from the tip of Gaspé Peninsula to the north end of the Green Mountain massif in Vermont, following the axis of the Green Mountain—Sutton Mountain—Shickshock Range anticlinorium, no known Precambrian rocks of about 1 b.y. or older occur. The southeastern limit of the 1 b.y. or older Precambrian rocks then follows the Green Mountain massif, the Berkshire massif, the Housatonic massif to the Hudson massif and thence southwest through north-central New Jersey to the Reading Prong (*see* Tilton and others, 1960; also references *in* Lyons and Faul, 1968). To the south and east of this belt, the only known basement rocks of 1 b.y. age are the gneisses of the Chester and related domes (Faul and others, 1963, p. 7) and the Baltimore Gneiss domes (Tilton and others, 1958; Wetherill and others, 1968).

Other Precambrian rocks southeast of this line are those of Avalon Peninsula in eastern Newfoundland, rock masses in Nova Scotia (including Cape Breton Island) and in New Brunswick, rock masses in eastern Massachusetts and adjacent Rhode Island, and rock masses near New York City (*see* Fig. 1). These masses are all considerably younger, in the range of 600 m.y. (Long, 1969; also references *in* Rodgers, 1967, 1968). Poole pointed out (1967, p. 15) that the platformal character of the Avalon platform was acquired only in latest Precambrian time. It seems reasonable to assume that the 1-b.y.-old rocks, coinciding roughly with the southeastern limit of the Taconide zone discussed in this paper, was approximately the margin of the lower Paleozoic craton, so that many of the orogenic processes described in this paper were processes that operated at an important paleotectonic junction. Rodgers (1968, p. 148) and Diment (1968, p. 407) suggested that this junction was the junction of continental and essentially oceanic crusts. The possible role of a junction of oceanic and continental crusts in orogeny already has been discussed by many authors, for example Dietz (1963), and Bird and Dewey (1970).

Note added in proof: See, in this connection, also the following two recent papers: 1. Hughes, C. J., and W. D. Brückner, 1971, Late Precambrian rocks of eastern Avalon Peninsula, Newfoundland—a volcanic island complex: Canadian Jour. Earth Sci. v. 8, p. 899-915. 2. Upadhyay, H. D., J. F. Dewey, and E.R.W. Neale, 1971, The Betts Cove ophiolite complex, Newfoundland: Appalachian oceanic crust and mantle: Geol. Assoc. Canada Proc. v. 24, p. 27-34. See also Church and Stevens, 1969.

Breakup of the Shelf

The stable configuration of paleogeographic elements in the pretectonic Taconides came to an end toward late Early Ordovician time, when the shelf area began to be broken up by high-angle faults. This is seen most clearly in the area surrounding the Taconic klippe, but similar events probably occurred elsewhere in the same zone; the early Paleozoic high-angle faults in the autochthon developed within the general span of late Early Ordovician to Middle Ordovician, but the age varied from one area to another.

Bathymetric changes resulted from this faulting; the areas of foundered blocks of the shelf became areas of deep water, but those of the uplifted blocks in part became areas of subaerial and submarine erosion or of nondeposition. As a whole, however, the amount of net uplift seems to have increased in a seaward direction at least as far as the region of deposition of the transitional facies; whereas in the shelf area, despite notable differential vertical movement, the over-all effect was one of subsidence.

Change of Sedimentary Regime

One effect of the fragmentation of the shelf and its partial foundering through block faulting was to alter completely the sedimentary pattern in the western part of the Taconide zone: the water depth was changed, and the bathymetric relation between the former shelf area and basin area was reversed. The source and nature of sediment supply and mode and direction of sediment transport were affected also. The sediments that had accumulated in the basin since the early Paleozoic were uplifted locally, to become sources of sediment supply and to produce second-cycle clastic material. The former areas of shallow-water carbonate accumulation became areas of deep water and accumulated clastic, argillaceous to quartzofeldspathic sediments which developed into the mantling black shale deposits. Thus, above this hiatus in sedimentary regime, represented in the stratigraphic column by a major regional unconformity (whose many oddities in the Taconic region have been discussed by Zen, 1967a, p. 40-44), is the record of the second major facies change described in a preceding section on Sedimentary Facies Changes — a regional facies change into black shale deposition that is vertical with time, rather than horizontal with lateral passage. This facies change, in fact, extends south into the southern Appalachians; the continuity of the facies into the latter area suggests similar geologic history that is yet to be deciphered.

Gravity Slides and Destruction of Sedimentary Facies Junction

Continued increase in relief between the former shelf area, now downdropped, and the former deep-water basin where the transitional facies was deposited and now elevated, finally led to the wholesale breakoff of the uplifted lower Paleozoic sedimentary cover of the former basin resulting in its west, northwest, or northward sliding. The earliest sedimentary record of this event is preserved at any given locality in the wildflysch-type conglomerate that is found in the mantling black shale sequence. These conglomerates have been described and interpreted by many people, especially in recent years: Brückner (1966), Tuke (1968), and Stevens (1968) for Newfoundland; Bailey and others (1928), Osborne (1956), and St-Julien (1968) for the Quebec City area; Hawley (1957) for northwestern Vermont; Zen (1961, 1967a) and Bird (1963, 1969) for the Taconic area; MacLachlan and Root (1966, p. 46), Platt and others (1969), and Alterman (1969) for southeastern Pennsylvania. These workers agree that the conglomerates, though of different ages in different places, indicate sources of sediments that were close by and had considerable relief. Recent workers consider that the source areas for the conglomerates were allochthonous masses, and that the culmination of the process of tectonically related sedimentation was the emplacement of enormous blocks of rocks of the basin facies into a sea of black shale deposition. The individual gravity slides are the end members of this series of increasingly coarse clasts.

Large-scale gravity sliding of rocks could lead to several initial geometric configurations relative to the original facies dispositions; these configurations, in turn, could generate different field relations when later eroded (Fig. 10). Initially the gravity slides could be completely severed, or could be at least apparently continuous with the autochthonous rocks in the hinterland. Upon subsequent erosion, an allochthonous mass could (1) be surrounded on all sides by rocks of a different facies (klippe); (2) remain apparently attached to the site of deposition, only the frontal part resting on rocks of a different facies across a movement surface (paraklippe); (3) be eroded back so that only a "stump" of the thrust sheet remains, permitting its interpretation as a thrust of small displacement. Which one of these situations obtains probably would depend on the vagaries of erosion, exposing the structures at different levels. I propose that the first situation is exemplified by the Humber Arm and the Taconic klippen as well as by the discontinuous allochthonous rocks of southeastern Pennsylvania. The

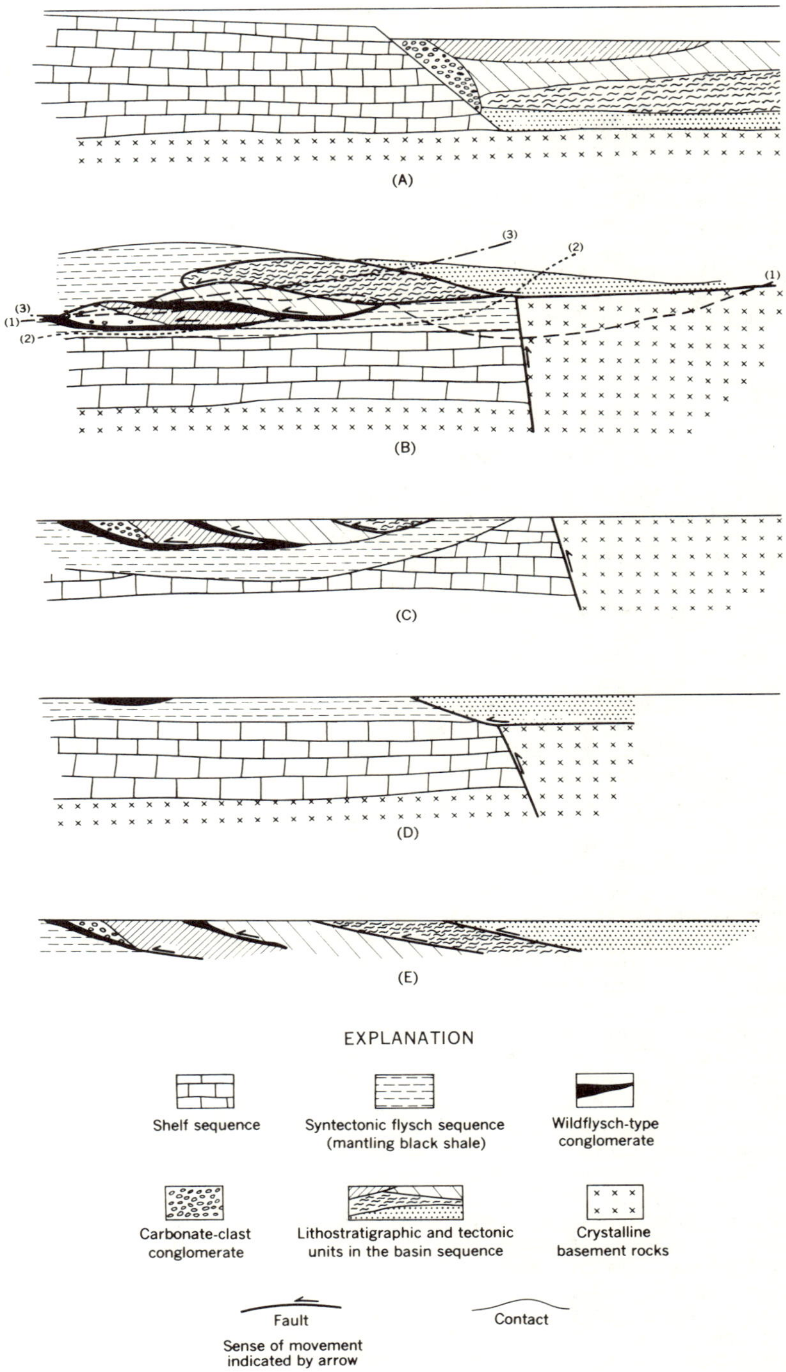
(A)
(3)
(2)
(1)
(3)
(1)
(2)
(B)
(C)
(D)
(E)
EXPLANATION
Shelf sequence
Syntectonic flysch sequence (mantling black shale)
Wildflysch-type conglomerate
Carbonate-clast conglomerate
Lithostratigraphic and tectonic units in the basin sequence
Crystalline basement rocks
Fault
Sense of movement indicated by arrow
Contact

Hare Bay allochthon probably is also a klippe. Some minor klippen also exist in the Quebec City area (Hubert, 1965; Pierre St-Julien, 1970, written commun.). The second situation is typified by much of the allochthonous masses of Quebec, directly south and east of Logan's Line, and possibly by the rocks of the Hare Bay allochthon; however, the precise relations are obscured either by imperfect geologic knowledge of the hinterland, by the cover of Silurian and Devonian rocks, or by the Atlantic Ocean. The third situation may be instanced by the relations at the Hinesburg thrust; here, the former existence of allochthonous masses of long travel is shown only by the wildflysch-type conglomerate on islands and the east shore of Lake Champlain just west of St. Albans, Vermont. Some areas in southeastern Pennsylvania and New Jersey also may prove to belong to this category.

One effect of gravity sliding was to conceal or destroy much of the junction between the lower Paleozoic shelf and basin facies. This junction is rarely preserved; Rodgers (1968) pointed to northwestern Vermont (*see also* Cady, 1969, p. 21) and southeastern Pennsylvania as two examples of preservation, and the area around White Bay in western Newfoundland may be another example (Heyl, 1937; E.R.W. Neale, 1969, oral commun.). For the greater part of the length of the Taconides, however, the nature of the junction can be studied only where the appropriate rocks have been accidentally preserved as tectonic slivers between or below slices of allochthons, or as blocks in wildflysch-type conglomerates in the mantling black shale (Zen and J. M. Bird, unpub. data). Such accidental records include carbonate beds, carbonate-clast conglomerates, and biohermal carbonate rocks, which are lithically foreign to the prevalent bedrock of the area, and which are indicative of specific environments of deposition otherwise no longer in evidence. For instance, the presence of carbonate-clast conglomerates in the allochthon as well as in the underlying wildflysch-type conglomerate in the Quebec City area shows clearly that a lower Paleozoic carbonate shelf area once existed here. Conversely, the lack of "brink"-type (that is, those de-

Figure 10. Diagrammatic sections illustrating the effect of erosion level on the apparent field relations, for the same set of initial geometric arrangement of allochthon and autochthon: (A), the presumed pretectonic relation; (B), the geometric relation after the emplacement of the slices of the allochthon according to the general scheme discussed in this paper; the three lines, 1, 2, and 3 label three possible erosion levels; (C), the observed field relations if erosion level is along line 1 of section (B) — the allochthon is obviously a klippe; (D), the observed field relations if erosion level is along line 2 of section (B) — the allochthon is vestigial, and the evidence for it consists of the wildflysch-type conglomerate and the "stump" which appears to be a simple thrust fault; (E), the observed field relations if erosion level is along line 3 of section (B) — a series of overlapping thrust sheets would be mapped without indication of the nature of the overridden rocks or the existence of a klippe. The klippen of Newfoundland, the Taconic klippe, and possibly the allochthon of New Jersey and Pennsylvania could correspond to section (C); the relations of northwestern Vermont and possibly part of the New Jersey and Pennsylvania area could correspond to section (D); the relations in the area south of Logan's Line in Quebec could correspond to section (E).

veloped at the shelf and basin junction; Zen and J. M. Bird, unpub. data) carbonate-clast conglomerates within the autochthon or the allochthon in the Taconic region suggests that part of the stratigraphic record may be lost.

The Room Problem

Discussions of the tectonic evolution of allochthons of the Taconide zone eventually evolve around the critical question of finding sufficient space in the root zone or depositional site to accommodate the travelled rocks. We will briefly review the question; a complete solution is nowhere at hand for any of the allochthonous masses. Our review will start with the Taconic klippe.

The Taconic klippe is, on the most conservative estimate, from 25 to 30 km (15 to 20 mi) wide, not taking into account any missing sedimentary facies, or the possibility that the nested slices originally might have been juxtaposed laterally rather than in vertical stratigraphic continuity (Zen, 1967a, p. 67). The probable depositional sites, the Precambrian massifs to the east, barely accommodate this minimum width, even with an apparently generous allowance for subsequent warping, erosion, and distributive shearing in the area of the proposed depositional sites; the problem appears most acute in northwestern Massachusetts, where the massifs plunge under lower Paleozoic sedimentary cover (Cady, 1969, p. 65; *see*, however, discussion *in* Zen, 1967a, p. 56-62). The rocks now overlying the eastern margin of the Green Mountain massif have been shoved westward by later distributive shears parallel to the bedding surfaces (Zen, 1967a, p. 57). The same situation may obtain on the east side of the Berkshire massif; it is possible, for instance, that the lithic changes within the lower Paleozoic eugeosynclinal sedimentary rocks east of the massif (described *by* Hatch and others, 1968, p. 179) are partly the result of major thrust faults that repeat as well as eliminate parts of the normal stratigraphic section. Such thrusts would be geometrically comparable to the Hinesburg fault in Vermont, except that rocks on both sides would be of the same general facies. The existence of such major dislocations seems reasonable by comparison with the relations at North Adams gap, just north of the Berkshire massif (Zen, 1967a, p. 58).

All along Logan's Line in Quebec, the allochthonous rocks pose a similar and severe room problem for their restoration. Except locally, however, the problem cannot be clearly defined yet, or its magnitude gauged, because the structures in the rocks are imperfectly known and because the possible site of original deposition has not been identified.

The room problem for the allochthonous rocks in the eastern part of Gaspé Peninsula (Riva, 1968, p. 1380) was recently analyzed by R. K. Stevens (1969, written commun.), who pointed out that if the pretectonic ultramafic intrusive rocks, such as that of Mt. Albert, are autochthonous, then an insuperable room problem arises for the traveled rocks immediately to the north. On the other hand, if the intrusive rocks are part of the allochthon, then the rocks for which a depositional site must be found becomes correspondingly more extensive. Because of the scale of the features and because of the total obliteration of the shelf which once must have existed, the distance of travel of the allochthonous masses must be measured in many if not many tens of kilometers.

Our information on the minimum travel of the allochthonous masses in the Kamouraska area is firmer than for the Gaspé Peninsula because Hubert (1967, Fig. 2) deciphered several nested thrust slices.

Palinspastic restoration of these slices would lead to a cumulative minimum movement on the order of 15 km (10 mi); the actual movement may be much more.

Therefore it seems clear that, whatever the geologic details, a loss of at least 15 to 30 km (10 to 20 mi) of supracrustal width of sedimentary rocks occurs along most of Logan's Line.

In Newfoundland, the room problem is severe, though here the distribution of sedimentary facies seems to indicate that the depositional site originally was west of the lower Paleozoic rocks now exposed east of the Cabot fault zone (Church and Stevens, 1969). This assignment is consistent with the suggestion of Rodgers and Neale (1963) for the depositional sites, earlier summarized. The allochthonous mass was estimated by Church and Stevens (1969) to be at least 50 km (30 mi) wide, however; the area west of the Cabot fault zone is insufficient to accommodate this amount. Here a "Verschluckung" type of crustal loss has been suggested (Church and Stevens, 1969; *see also* Append. 5).

At the other end of the Taconide zone, the problem of restoration of the allochthonous masses of southeastern Pennsylvania to a site of deposition is also a problem. As stated before, Kay (1941) early suggested that the Martic Line in southeastern Pennsylvania is the dislocation along which the allochthonous rocks were derived. If so, then a shortening of the supracrustal section would be indicated, though of uncertain amount and mechanism.

Later Taconic Orogenic Events

In the preceding pages, we have outlined the sedimentary and tectonic patterns over a strip of land a few tens of kilometers wide and a few thousand kilometers long. We have followed a possible causal sequence of events from the beginning of the Paleozoic to late Middle Ordovician time, when the gravity slides began to peel off from the sites of deposition. We have seen that this inception of gravity sliding was not simultaneous throughout the Taconides; rather, it varied from place to place over a span of time ranging apparently from latest Early Ordovician in Newfoundland to latest Trenton in northwestern Vermont.

As might be expected, Taconic tectonic events did not stop suddenly with the gravity slides. The later activities are less well known. The events around and southwest of the Taconic klippe will be briefly summarized here in the hope that it may be used as a guide in future research.

As previously pointed out (Zen, 1967b), the upper (younger) slices of the Taconic klippe apparently were emplaced under a different set of physical conditions than were the lower (older) slices. The lower slices are rich in slump structures and soft-rock deformation, which appear to be absent in the upper slices. The lower slices comprise lowest Cambrian to Middle Ordovician rocks, but the upper slices appear to contain only Precambrian to lowest Cambrian rocks, so that the two groups might show no stratigraphic overlap (Zen, 1967a, p. 15). The lower slices have associated wildflysch-type conglomerate and few carbonate slivers. The upper slices have no wildflysch-type conglomerate, but have many carbonate slivers; these slivers are definitely not sedimentary conglomerates (Zen and Ratcliffe, 1966, 1971), but are associated with mylonite as well as intensely sheared rocks, indicating brittle deformation. Emplacement of the lower slices is not known to have caused penetrative deformation in the underlying autochthon, but the emplacement of the upper slices in southwestern Massachusetts and adjacent Connecticut has been associated with isoclinal recumbent folding of the autochthon over which the slices moved (Zen and Ratcliffe, 1971; Zen, 1969).

These sharply contrasting types of behavior indicate that the lower and upper slices of the Taconic klippe were not emplaced under similar tectonic regimes. Recent studies (for example, Zen, 1969; Ratcliffe, 1965, unpub. data) indicate that some of the rocks in the southeastern part of the Taconic klippe have been polymetamorphosed. Study of the relations among the various tectonic, metamorphic, and sedimentary events in this area indicate that the sequence, from older to younger, was: high-angle faults; mid-Ordovician unconformity; emplacement of earlier gravity slides (not preserved here); emplacement of later slices of the allochthon with concomitant recumbent folding of both autochthon and allochthon; regional low-grade (biotite grade or lower) metamorphism; possible burial by tectonic pile-up or sedimentation; later (Acadian) metamorphism and deformation.

Zen (1969) and Ratcliffe (1965, 1969b) interpreted the early metamorphic event affecting Middle Ordovician and older rocks in the southern Taconic region, and Ratcliffe (1968, 1970) also interpreted the same effect in the area between Dutchess County and New York City, including the area near Peekskill, New York, as a Taconic event; this could be Late Ordovician or Early Silurian in age. Radiometric data indicate that the event may be regionally widespread. Thus, muscovite that grew in the plane of early metamorphic foliation along the Sutton Mountain anticlinorial axis in Quebec has yielded ages of 420-440 ± m.y. (Rickard, 1965). Similar metamorphic ages were determined for rocks from northwestern Vermont (Cady, 1969, p. 160, 163; U.S. Geological Survey, unpub. data) and from the New York City area (Long and Kulp, 1962). Because the emplacement of the highest slice of the Taconic allochthon in southwestern Massachusetts predates this metamorphic event, Zen (1969) concluded that the emplacement itself was Late Ordovician or earliest Silurian in age; this emplacement caused isoclinal recumbent folding in the underlying autochthon.

Gray and his co-workers (1958) deciphered recumbent folding of the pre-Middle Ordovician carbonate rocks of the shelf sequence (*see also* MacLachlan and Root, 1966) in the Lebanon, Pennsylvania, area west of the Precambrian rocks of the Reading Prong. Recumbent folds have been deciphered in the Cambrian and Lower and Middle Ordovician carbonate rocks in the area farther to the northeast and continuing into New Jersey, north of the Precambrian rocks of the New Jersey Highlands (*see* Sherwood, 1961; also especially Drake, 1969, for description and references). Both systems of recumbent folds are likely Ordovician in age (*see*, for example, MacLachlan and Root, 1966, p. 35, 46; Drake, 1969, p. 115).

This last episode of deep-seated thrusting may include the thrusts that displaced the western front of the Berkshire massif (Ratcliffe, 1969a). The thrusting of the Precambrian crystalline rocks of the New Jersey Highlands and the Reading Prong (geometrically a continuation of the Hudson massif) on the Musconetcong fault was interpreted as a late Taconic event (Drake, 1969, p. 92, 115). This thrusting, indeed, could have caused the recumbent folding of the carbonate rocks described in the preceding paragraph. Isachsen (1964), in a review of the basement rocks of the eastern United States, suggested that the entire Hudson massif might be allochthonous; that the northwestern margin is a thrust surface has been borne out by detailed mapping of outliers of these rocks north of the main body (Jaffe and Jaffe, 1967). Ratcliffe (1970), however, concluded that the northern end of the Hudson massif is more likely parautochthonous; it is possible that the amount of horizontal transport increased southwestward toward Reading, Pennsylvania.

Dating of movement of these Precambrian and associated younger rocks could be inferred from another argument. The Green Pond outlier of predominantly Silurian to Middle Devonian rocks rests on the Precambrian of the Hudson massif near the New York–New Jersey line. These rocks are little deformed and unmetamorphosed (*see* Kümmel and Weller, 1902); also, except for a major break between the Shawangunk Formation (Silurian) and the underlying Ordovician(?) black shale, and possibly another, lesser break about the level of the Oriskany Sandstone (Lower Devonian; Cooper, 1942), these rocks show no internal hiatus of importance. The Silurian and Devonian rocks lithically match autochthonous rocks of this age in the Catskill–Pocono Plateau area (Fig. 1), as well as in the Becraft Mountain outlier in the Taconic region (Fig. 7). This fact, and the fact that the Devonian rocks in the Green Pond outlier span the Acadian orogeny (Boucot, 1968, p. 92), would seem to exclude an interpretation that the older parts of the Silurian and Devonian rocks were deposited in an area of the basin farther to the east and were thrust in during the early, premetamorphic phase of the Acadian orogeny; these rocks almost certainly are autochthonous. The tectonic movement of the Precambrian basement rocks, therefore, probably predated the Shawangunk Formation (Silurian). Earlier it has been stated that the movement of the Precambrian crystalline rocks is later than Normanskill because of the relations at Jutland, New Jersey. The dating, Late Ordovician or possibly Early Silurian, agrees well with the conclusions of Drake (1969).

The presence of Cambrian and Ordovician rocks of the shelf sequence in the Green Pond outlier (Kümmel and Weller, 1902) is not a compelling argument against the allochthony of the Precambrian rocks because synchronous rocks of the shelf sequence now occur even farther to the south and east (for example, the New York City area; Hall, 1968a).

Fisher (1969) described a facies of rocks on the east side of the Catskill plateau which could be Upper Ordovician and which was interpreted as a post-tectonic "molasse" facies. This interpretation accords broadly with the chronology offered in this paper.

A deep-seated environment for the later phases of the Taconic orogeny can be inferred also from the environment of contact metamorphism of Middle Ordovician rocks by the Cortlandt Complex near Peekskill, New York. The Ordovician rocks are of the mantling shale sequence which was folded and metamorphosed before the intrusion of the mafic plutons at Cortlandt (Ratcliffe, 1968) about 435 m.y. ago (based on a K-Ar age determination of single biotite; Long and Kulp, 1962). The contact metamorphic rocks in one quarry (Friedman, 1956, p. 28) carried kyanite, andalusite, and sillimanite, indicating a temperature-pressure environment of metamorphism near the triple point of the aluminum silicates, about 620° C and 5.5 kb (Richardson and others, 1969). The implied depth of burial is about 18 km. Because of the small age difference between the intrusive rock and the country rock, a rapid rate of burial, on the order of 1 mm/yr, is indicated. Rosenfeld (1968, p. 196) suggested from a study of rotation of almandine porphyroblasts in lower Paleozoic rocks the existence of pre-Silurian almandine-grade metamorphism in southeastern Vermont. The movement implied by the almandine rotation could have been related to the tectonic pile-up implied by the metamorphic conditions at the Peekskill area; perhaps the large thrust faults in the North Adams, Massachusetts, area, summarized and inter-

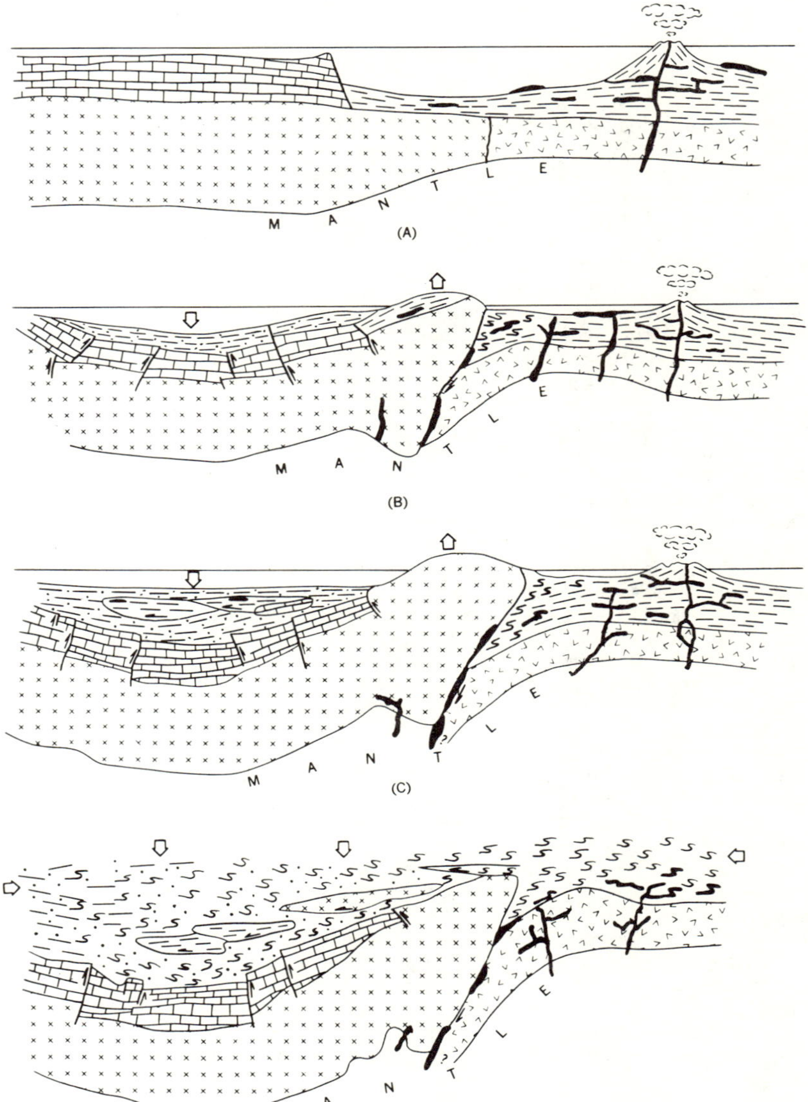

Figure 11. Schematic cross sections representing the stages of evolution of the Taconides during the Taconic orogeny. (A), pretectonic conditions: deposition of a shelf sequence was succeeded seaward by deposition of a basin sequence which included volcanic rocks. This facies transition occurred on the continental crust, but near the junction between a thick continental and a thin oceanic crust. (B), early stage of deformation: the oceanic crust began to flow under the continental crust. The edge of the continental crust was rafted up owing to its lesser density, but the more interior region of the continent was downwarped. This led to a reversal of the pretectonic bathymetry, and

preted (Zen, 1967a, p. 60) as covering up much of the depositional area of the Taconic rocks, were also a Late Ordovician instead of an Acadian event (*see also* Ratcliffe, 1969b). If so, the geographic extent of the Taconides would have to be vastly expanded from the narrow strip discussed in this paper.

Speculations on the Underlying Mechanism for Tectonism

In the preceding pages we have attempted to show that the orogenic processes in the Taconides — the fragmentation of the shelf-platform, the inception of the black shale facies of sedimentation, the emplacement of the gravity slides — are everywhere chronologically mutually connected events, and that they also can be causally related by a simple mechanism of bathymetric reversal. What caused the reversal, however, has not been mentioned (*see also* Append. 3).

Because of the close spatial relation between the gravity ridge and trough and the major paleotectonic units in the Taconides, I propose that they are causally connected; whatever caused the gravity anomalies likely gave rise to the crustal disturbances leading to the reversal of bathymetry.

Interpreting the gravity relations around the Taconic klippe, I previously suggested (Zen, 1967a, p. 72) that the common denominator could be a lateral transfer of material in the upper mantle or lower crust, from the area underlying the lower Paleozoic shelf to the area underlying the axial region of deep water. The region-by-region review of the gravity anomalies and their permissible interpretation, given

---

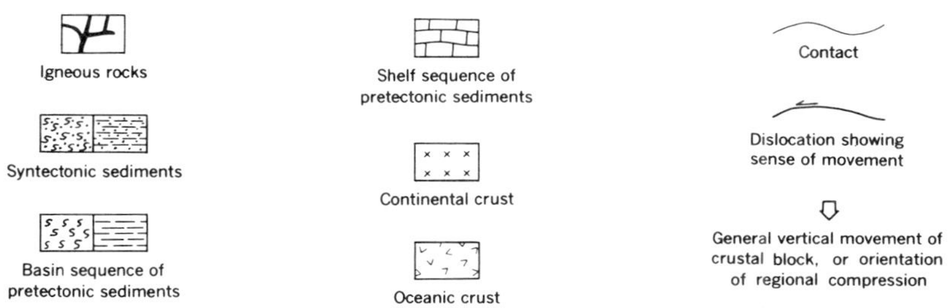

the former shelf area became the depositional basin for syntectonic sediments (flysch) partly generated from the former basin. Igneous material, at least partly generated from the upper mantle, was introduced into the crust along fracture zones, and was especially abundant in the thinner oceanic crustal region. The igneous rocks were both intrusive and extrusive. A zone of gravity high was created along the zone of underflow because excess matter was accumulated here. (C), a later stage of development of the process: wholesale sliding of basin sediments formerly deposited along the edge of the continental block had occurred; these were incorporated in the syntectonic flysch sequence as allochthons. Crumpling of sediments along the zone of crustal underflow continued. (D), the entire area had been buried under thick piles of rocks through both deposition and tectonism. Metamorphism and deep-seated deformation continued. Part of the crystalline continental crust itself was thrust over the allochthonous sedimentary material. In all the sections, the left-hand end is toward the north and west.

earlier, shows that the geologic and geophysical relations obtaining in the Taconic area are true for the Taconides as a whole, so a common explanation is called for. A reasonable explanation presumably should also recognize the fact that the process occurred along the periphery of the 1-b.y.-old craton.

I now suggest that the common underlying cause of the tectonic processes was not a lateral transfer of material, but a movement of the oceanic crustal plate north and west toward the cratonal plate (Fig. 11; *see also* Dewey, 1969; Bird and Dewey, 1970, for somewhat different reconstructions). Because the cratonal plate was lower in density, the edge of the craton was rafted up by the addition of oceanic material underneath. This led to the uplift of the edge of the margin of the 1-b.y.-old craton. Farther to the north and west, the craton became downwarped by compression caused by the impinging oceanic plate; because deformation was in the near-surface brittle regime, the shelf rocks and the upper crust broke by fracture, and longitudinal faults having predominantly dip-slip movement sense resulted. The uplift of the edge of the craton and the downdrop of the adjacent former shelf area provided the bathymetric reversal indicated by the stratigraphic record of the Middle Ordovician rocks in the Taconides.

The movement of the oceanic crust under the craton generated a tectonically mobile belt and provided opportunity for intrusion and extrusion of dense, mafic, and ultramafic rocks. This activity can be expected to be especially concentrated near the junction of the two crustal plates, thus accounting for the location of the belt of ultramafic intrusive rocks near the limit of the craton, and for the presence of the ultramafic sheets in the allochthons of Newfoundland, interpreted by Church and Stevens (1969) as an ophiolite sequence that poured out onto the sea floor. Finally, this scheme could account for the chain of volcanic islands in the eugeosynclinal part of the Ordovician basin (Kay, 1967a; Thompson and others, 1968; Naylor, 1968; Berry, 1968).

The addition of dense material directly beneath the cratonal margin explains the presence of a belt of positive Bouguer anomaly coincident with the edge of the craton.

Continued compression led to the deep burial of the Ordovician and older sedimentary rocks; hence the later phases of the Taconic orogeny were deep seated, and, locally, as in the New Jersey and Pennsylvania area, such later deformation involved thrusting of the cratonal rocks of the rafted margin themselves. Although the very distinction between craton and mobile belt became blurred when this stage was reached, by and large throughout the Taconic orogeny the diastrophic process appears to have concentrated along the boundary zone between the craton and the mobile belt, and the processes that operated on either side of this boundary were different: predominantly thin-skinned deformation occurred on the craton, not only during the Taconic orogeny, but in subsequent episodes of deformation as well (for instance, the thin-skinned tectonics of the Valley and Ridge province all occurred on the craton; *see* Rodgers, 1963; Gwinn, 1964); predominantly thick-skinned tectonics occurred in the mobile belt.

The Taconic orogeny was the first major episode of deformation affecting the lower Paleozoic sedimentary rocks of the Appalachian area. This orogeny was not as intense as the Acadian orogeny in the northern Appalachian region, in that the latter orogeny involved greater area, led to more and larger scale development of nappes and deep-seated thrusts, produced more widespread and higher grades of regional metamorphism, and gave rise to larger scale plutonism. However, the later Acadian orogeny seems to have followed the tectonic trend set by the Taconic orogeny, and was not as productive of ultramafic intrusive bodies as was the Taconic orogeny (*see*, in this connection, the discussion *by* Hess, 1955, p. 395). In this sense, the Taconic orogeny served as the template for the evolution of the northern Appalachian orogenic system, its effects profoundly influencing those of subsequent events.

Our story of the evolution of the Taconide zone comes to a close. It is a highly imperfect story; many problems, big and small, remain unresolved, and many blank areas remain in the picture. It is as if the geologist is made to put together a jigsaw puzzle of uncertain design, for which the pieces are not guaranteed to interlock, and for which many pieces are known to be missing, others may be missing, and most have their colors and patterns blurred by hard use. Extra pieces might also have been thrown in, unbeknownst to the player. Despite these real or artificial difficulties, however, studies of the past several decades do permit a general tectonic and sedimentary pattern to be discerned. In this paper I have traced some of these patterns and attempted to fill in the gaps. How successful this effort has been, future research will find out.

References Cited

Numbers in square brackets are page numbers on which the reference is cited.

Alterman, I. B., *1969, An Ordovician boulder conglomerate at the base of the Hamburg klippe, east-central Pennsylvania [abstr.]: Geol. Soc. America Abstr. with Programs for 1969, pt. 1, p. 1-2.* [29, 41, fig. 2]

Bailey, E. B., Collet, L. W., and Field, R. M., *1928, Paleozoic submarine landslips near Quebec City: J. Geol., v. 36, p. 577-614.* [34, 41]

Barnett, S. G., 3d, *1965, Conodonts of the Jacksonburg Limestone (Middle Ordovician) of northern New Jersey and eastern Pennsylvania: Micropaleontology, v. 11, p. 59-80.* [31]

Bayley, W. S., Salisbury, R. D., and Kümmel, H. B., *1914, Description of the Raritan quadrangle [New Jersey]: U.S. Geol. Survey Geol. Atlas, Folio 191, 32 p.* [31, fig. 2]

Behre, C. H., Jr., *1933, Slate in Pennsylvania: Pennsylvania Geol. Survey 4th ser. Bull. M-16, 400 p.* [12]

Belt, E. S., *1968, Post-Acadian rifts and related facies, eastern Canada, in Zen, E-an, White, W. S., Hadley, J. B., and Thompson, J. B., Jr., eds., Studies of Appalachian geology, northern and maritime: New York, Interscience Publishers, p. 95-116.* [26]

Berry, W.B.N., *1962, On the Magog, Quebec, graptolites: Am. J. Sci., v. 260, p. 142-148.* [16]

— *1968, Ordovician paleogeography of New England and adjacent areas based on graptolites, in Zen, E-an, White, W. S., Hadley, J. B., and Thompson, J. B., Jr., eds., Studies of Appalachian geology, northern and maritime: New York, Interscience Publishers, p. 23-34.* [17, 39, 50]

— *1970, Review of late Middle Ordovician graptolites in eastern New York and Pennsylvania: Am. J. Sci., v. 269, p. 304-313.* [27, fig. 2]

Betz, Frederick, Jr., *1939, Geology and mineral deposits of the Canada Bay area, northern Newfoundland: Newfoundland Geol. Survey Bull. 16, 53 p.* [23, 25, 26, fig. 2]

Bird, J. M., *1963, Sedimentary structures in the Taconic sequence rocks of the southern Taconic region, in Bird, J. M., ed., Stratigraphy, structure, sedimentation and paleontology of the southern Taconic region, eastern New York — Field trip 3, Albany, 1963, guidebook: New York, Geol. Soc. America, p. 5-21.* [41]

— *1969, Middle Ordovician gravity sliding in the Taconic region, in Kay, Marshall, ed., North Atlantic — geology and continental drift: Am. Assoc. Petroleum Geologists Mem. 12, p. 670-686.* [27, 41, fig. 2]

Bird, J. M., and Dewey, J. F., *1970, Lithosphere plate: Continental margin tectonics and the evolution of the Appalachian orogen: Geol. Soc. America Bull., v. 81, p. 1031-1060.* [40, 50]

Bird, J. M., and Rasetti, Franco, *1968, Lower, Middle and Upper Cambrian faunas in the Taconic sequence of eastern New York: Stratigraphic and biostratigraphic significance: Geol. Soc. America Spec. Paper 113, 66 p.* [27]

Black, R. F., *1964, Palaeomagnetic support of the theory of rotation of the western part of the island of Newfoundland: Nature, v. 202, p. 945-948.* [71]

Boucot, A. J., *1968, Silurian and Devonian of the northern Appalachians, in Zen, E-an, White, W. S., Hadley, J. B., and Thompson, J. B., Jr., eds., Studies of Appalachian geology, northern and maritime: New York, Interscience Publishers, p. 83-94.* [47]

Bricker, Owen, *1960, Ordovician volcanics of the Bunker Hills, Lebanon County, Pennsylvania, in Wise, D. U., and Kauffman, M. E., eds., Some tectonic and structural problems of the Appalachian Piedmont along the Susquehanna River: Field Conf. Pennsylvania Geologists, 25th, Lancaster, 1960, p. 92-99.* [29]

Broughton, J. G., Fisher, D. W., Isachsen, Y. W., Rickard, L. V., and Offield, T. W., *Compilers and eds., 1962, Geologic map of New York, 1961: New York State Mus. and Sci. Service, Geol. Survey, Map and Chart Ser., no. 5, 5 sheets, scale 1:250,000, text, 42 p.* [17]

Brückner, W. D., *1966, Stratigraphy and structure of west-central Newfoundland, in Poole, W. H., ed., Geology of parts of Atlantic provinces — Geol. Assoc. Canada and Mineralog. Assoc. Canada, 1966, Guidebook: Halifax, Nova Scotia, Nova Scotia Dept. Mines, p. 137-155.* [25, 26, 39, 41, fig. 2]

Cady, W. M., *1945, Stratigraphy and structure of west-central Vermont: Geol. Soc. America Bull., v. 56, p. 515-587.* [20]

— *1960, Stratigraphic and geotectonic relationships in northern Vermont and southern Quebec: Geol. Soc. America Bull., v. 71, p. 531-576.* [16, 17, 19, 20, 35, fig. 5, fig. 2, fig. 9]

— *1967, Geosynclinal setting of the Appalachian Mountains in southern Quebec and northwestern New England, in Clark, T. H., ed., Appalachian tectonics: Royal Soc. Canada Spec. Pub. 10, p. 57-68.* [11]

— *1968a, The lateral transition from the miogeosynclinal to the eugeosynclinal zone in northwestern New England and adjacent Quebec, in Zen, E-an, White, W. S., Hadley, J. B., and Thompson, J. B., Jr., eds., Studies of Appalachian geology, northern and maritime: New York, Interscience Publishers, p. 151-162.* [12, 16, 18, 19]

— *1968b, Tectonic setting and mechanism of the Taconic slide: Am. J. Sci., v. 266, p. 563-578.* [19, 69]

— *1969, Regional tectonic synthesis of northwestern New England and adjacent Quebec: Geol. Soc. America Mem. 120, 181 p.* [19, 43, 44, 46]

Carswell, L. D., Hollowell, J. R., and Platt, L. B., *1968, Geology and hydrology of the Martinsburg Formation in Dauphin County, Pennsylvania: Pennsylvania Geol. Survey 4th ser. Bull. W-24, 54 p.* [29]

Chidester, A. H., *1968, Evolution of the ultramafic complexes of northwestern New England, in Zen, E-an, White, W. S., Hadley, J. B., and Thompson, J. B., Jr., eds., Studies of Appalachian geology, northern and maritime: New York, Interscience Publishers, p. 343-354.* [16]

Church, W. R., *1969, Metamorphic rocks of the Burlington Peninsula and adjoining areas of Newfoundland, and their bearing on continental drift in the North Atlantic, in Kay, Marshall, ed., North Atlantic — geology and continental drift: Am. Assoc. Petroleum Geologists Mem. 12, p. 212-233.* [17]

Church, W. R., and Stevens, R. K., *1969, Crustal evolution of the western margin of the Newfoundland Appalachians [abstr.]: Geol. Soc. America Spec. Paper 121, p. 53-54.* [45, 50]

Clark, G. S., and Kulp, J. L., *1968, Isotopic age study of metamorphism and intrusion in western Connecticut and southeastern New York: Am. J. Sci., v. 266, p. 865-894.* [9]

Clark, T. H., *1951, New light on Logan's Line: Royal Soc. Canada Trans., v. 45, ser. 3, sect. 4, p. 11-22.* [33]

— *1964a, Upton area: Quebec Dept. Nat. Resources Geol. Rept. 100, 37 p.* [35]

— *1964b, St. Hyacinthe area (west half): Quebec Dept. Nat. Resources Geol. Rept. 101, 128 p.* [35]

Clark, T. H., and Eakins, P. R., *1968, The stratigraphy and structure of the Sutton area of southern Quebec, in Zen, E-an, White, W. S., Hadley, J. B., and Thompson, J. B., Jr., eds., Studies of Appalachian geology, northern and maritime: New York, Interscience Publishers, p. 163-176.* [35, fig. 2]

Cleaves, E. T., Edwards, Jonathan, Jr., and Glaser, J. D., *Compilers and eds., 1968, Geologic map of Maryland: Maryland Geol. Survey, scale 1:250,000.* [fig. 8]

Cooper, G. A., *chairman, 1942, Correlation of the Devonian sedimentary formations of North America [Chart no. 4]: Geol. Soc. America Bull., v. 53, p. 1729-1794.* [47]

Cooper, J. R., *1937, Geology and mineral deposits of the Hare Bay area: Newfoundland Dept. Nat. Resources, Geol. Sec. Bull. 9, 36 p.* [23, 26]

Cox, Allan, and Doell, R. R., *1960, Review of paleomagnetism: Geol. Soc. America Bull., v. 71, p. 645-768.* [67]

Crowley, W. P., *1969, Stratigraphic evidence for a volcanic origin of part of the Bel Air belt of Baltimore Gabbro Complex in Baltimore County, Maryland [abstr.]: Geol. Soc. America Abstr. and Programs for 1969, pt. 1, p. 10.* [33]

Cumming, L. M., *1967, Platform and klippe tectonics of western Newfoundland — A review, in Clark, T. H., ed., Appalachian tectonics: Royal Soc. Canada Spec. Pub. 10, p. 10-17.* [25, 26]

Currier, L. W., and Jahns, R. H., *1941, Ordovician stratigraphy of central Vermont: Geol. Soc. America Bull., v. 52, p. 1487-1512.* [16]

Dewey, J. F., *1969, Evolution of the Appalachian/Caledonian orogen: Nature, v. 222, p. 124-129.* [50]

Dietz, R. S., *1963, Collapsing continental rises: an actualistic concept of geosynclines and mountain building: Jour. Geology, v. 71, p. 314-333.* [40]

Diment, W. H., *1968, Gravity anomalies in northwestern New England, in*

Zen, E-an, White, W. S., Hadley, J. B., and Thompson, J. B., Jr., eds., Studies of Appalachian geology, northern and maritime: New York, Interscience Publishers, p. 399-414. [18, 20, 29, 37, 40, fig. 7]

Doll, C. G., Cady, W. M., Thompson, J. B., Jr., and Billings, M. P., *Compilers and eds., 1961, Centennial geologic map of Vermont: Vermont Geol. Survey, scale 1:250,000.* [16, 17, 19, fig. 4, fig. 2, fig. 5]

Dominion Observatory Ottawa, *1967, Bouguer gravity map of Canada: Canada Dept. Energy, Mines and Resources, Observatories Br., 4 sheets, scale 1:2,534,400.* [18, 37, fig. 3]

Drake, A. A., Jr., *1969, Precambrian and lower Paleozoic geology of the Delaware Valley, New Jersey—Pennsylvania, in Seymour Subitsky, ed., Geology of selected areas in New Jersey and eastern Pennsylvania and guidebook of excursions: New Brunswick, New Jersey, Rutgers Univ. Press, p. 51-131.* [12, 29, 31, 33, 46, 47, fig. 2]

Dresser, J. A., and Denis, T. C., *1944, Geology of Quebec; V. 2, Descriptive geology: Quebec Dept. Mines Geol. Rept. 20, 544 p.* [68, fig. 2, fig. 9]

Elam, J. G., *1960, Geology of the Troy South and East Greenbush quadrangles, New York: Rensselaer Polytech. Inst., Ph.D. thesis, 232 p. (available on microfilms or by interlibrary loan).* [11, 69]

Emerson, B. K., *1917, Geology of Massachusetts and Rhode Island: U.S. Geol. Survey Bull. 597, 289 p.* [17]

Emmons, Ebenezer, *1841, Geology of the Montmorenci: Am. Mag., v. 1, p. 146-150 (reprinted, Am. Geologist, v. 2, p. 94-100, 1888).* [34]

Enos, Paul, *1969, Cloridorme Formation, Middle Ordovician flysch, northern Gaspé Peninsula, Quebec: Geol. Soc. America Spec. Paper 117, 66 p.* [12, 13, 16, 17, 33, fig. 2]

Faul, Henry, Stern, T. W., Thomas, H. H., and Elmore, P.L.D., *1963, Ages of intrusion and metamorphism in the northern Appalachians: Am. J. Sci., v. 216, p. 1-19.* [39]

Fisher, D. W., *1962, Correlation of the Ordovician rocks in New York State: New York State Mus. and Sci. Service, Geol. Survey Map and Chart ser., no. 3.* [17]

— *1969, Quinquallochthonous succession and a new molasse in the southern Hudson Valley and their bearing on New York tectonic history* [*abstr.*]*: Geol. Soc. America Abstracts with Programs for 1969, pt. 7, p. 66.* [47]

Friedman, G. M., *1956, The origin of spinel-emery deposits, with particular reference to those of the Cortlandt Complex, New York: New York State Mus. Bull. 351, 68 p.* [47]

Gates, R. M., and Christensen, N. I., *1965, Bedrock geology of the West Torrington quadrangle: Connecticut State Geol. and Nat. History Survey Quad. Rept. 17, 38 p.* [17]

Goodacre, A. K., Brule, B. G., and Cooper, R. V., *1969, Results of regional underwater gravity surveys in the Gulf of St. Lawrence, with map: Ottawa, Dominion Observatory, 24 p.* [26, 67, fig. 3]

Gray, Carlyle, Geyer, A. R., and McLaughlin, D. B., *1958, Geologic map of the Richland quadrangle, Pennsylvania: Pennsylvania Geol. Survey, 4th ser., Atlas 167D, scale 1:24,000.* [46]

Gray, Carlyle, and others, *1960, Geologic map of Pennsylvania: Pennsylvania Geol. Survey, 4th ser., 2 sheets, scale 1:250,000.* [17, 29, fig. 2, fig. 8]

Gwinn, V. E., *1964, Thin-skinned tectonics in the Plateau and northwestern Valley and Ridge provinces of the central Appalachians: Geol. Soc. America Bull., v. 75, p. 863-900.* [50]

Hall, L. M., *1968a, Bedrock geology in the vicinity of White Plains, New York, Trip A, in Finks, R. M., ed., Guidebook to field excursions: New York State Geol. Assoc., 40th Ann. Mtg., Flushing, New York, 1968, p. 7-31.* [13, 47]

— *1968b, Geology in the Glenville area, southwesternmost Connecticut and southeastern New York, Trip D-6, in Orville, P. M., ed., New England Intercollegiate Geol. Conf., 60th Ann. Mtg., Oct. 1968, Guidebook for fieldtrips in Connecticut: Connecticut State Geol. and Nat. History Survey Guidebook 2, p. D6-1 - D6-12.* [17]

Harper, C. T., *1968, Isotopic ages from the Appalachians and their tectonic significance: Canadian J. Earth Sci., v. 5, p. 49-59.* [9]

Hatch, N. L., Jr., Schnabel, R. W., and Norton, S. A., *1968, Stratigraphy and correlation of the rocks on the east limb of the Berkshire anticlinorium in western Massachusetts and north-central Connecticut, in Zen, E-an, White, W. S., Hadley, J. B., and Thompson, J. B., Jr., eds., Studies of Appalachian geology, northern and maritime: New York, Interscience Publishers, p. 177-184.* [44]

Hawley, David, *1957, Ordovician shales and submarine slide breccias of northern Champlain valley in Vermont: Geol. Soc. America Bull., v. 68, p. 55-94.* [19, 41, fig. 2, fig. 4]

— *1969, Sedimentary characteristics and tectonic deformation of Middle Ordovician shales in northwestern Vermont north of Malletts Bay, in Barnett, S. G., 3d, ed., Guidebook to field excursions: New York State Geol. Assoc., 41st Ann. Mtg., Plattsburgh, New York, 1969, p. 6-20.* [13]

Hess, H. H., *1955, Serpentinites, orogeny, and epeirogeny, in Poldervaart, Arie, ed., Crust of the earth — A symposium: Geol. Soc. America Spec. Paper 62, p. 391-407.* [51]

Heyl, G. R., *1937, The geology of the Sops Arm area, White Bay, Newfoundland: Newfoundland Dept. Nat. Resources Bull. 8, 42 p.* [26, 43]

Hopson, C. A., *1964, The crystalline rocks of Howard and Montgomery Counties, in The geology of Howard and Montgomery Counties: Baltimore, Maryland Geol. Survey, p. 27-215.* [31, 33]

Horne, G. S., *1969, Early Ordovician chaotic deposits in the central volcanic belt of northeastern Newfoundland: Geol. Soc. America Bull., v. 80, p. 2451-2464.* [17]

Hubert, Claude, *1965, Stratigraphy of the Quebec Complex in the L'Islet—Kamouraska area, Quebec: McGill Univ., Ph.D. thesis, 169 p. (available on microfilm or by interlibrary loan).* [33, 34, 36, 39, 43, fig. 2]

— *1967, Tectonics of part of the Sillery Formation in the Chaudière—Matapédia segment of the Québec Appalachians, in Clark, T. H., ed., Appalachian tectonics: Royal Soc. Canada Spec. Pub. 10, p. 33-40.* [33, 34, 44, fig. 2, fig. 9]

Hubert, Claude, Lajoie, J., and Léonard, M. A., *1970, Deep sea sediments in the lower Paleozoic Québec Supergroup, in Lajoie, J., ed., Flysch sedimentology in North America: Geol. Assoc. Canada Special Paper 7, p. 103-125.* [34]

Innes, M.J.S., and Argun-Weston, A., *1967, Gravity measurements in Appalachia and their structural implications, in Clark, T. H., ed., Appalachian tectonics: Royal Soc. Canada Spec. Pub. 10, p. 69-83.* [18, 36, 37]

Isachsen, Y. W., *1964, Extent and configuration of the Precambrian in northeastern United States: New York Acad. Sci. Trans., ser. 2, v. 26, p. 812-829.* [46]

Jaffe, H. W., and Jaffe, E. B., *1967, Structure and petrology of the Precambrian allochthon and Paleozoic sediments of the Monroe area, New York, in Waines, R. H., ed., Guidebook to field trips: New York State Geol. Assoc. 39th Ann. Mtg., New Paltz, 1967, p. F1-F17.* [46]

Johnson, Helgi, *1941, Paleozoic lowlands of northwestern Newfoundland: New York Acad. Sci. Trans., ser. 2, v. 3, no. 6, p. 141-145.* [26]

Jonas, A. I., and Stose, G. W., *1930, Lancaster quadrangle, geology and mineral resources: Atlas of Pennsylvania, no. 168: Pa. Topographic and Geologic Survey, Harrisburg, Pennsylvania, 106 p.* [31]

Kauffman, M. E., and Campbell, Lyle, *1969, Revised interpretation of the Cambrian Kinzers Formation in southeastern Pennsylvania [abstr.]: Geol. Soc. America Abstracts with Programs for 1969, pt. 1, p. 32-33.* [31]

Kay, Marshall, *1935, Distribution of Ordovician altered volcanic materials and related clays: Geol. Soc. America Bull., v. 46, p. 225-244.* [17]

— *1941, Taconic allochthons and the Martic thrust: Science, v. 94, p. 73.* [29, 31, 45]

— *1945, Paleogeographic and palinspastic maps: Am. Assoc. Petroleum Geologists Bull., v. 29, p. 426-450.* [23]

— *1966, Comparison of the lower Paleozoic volcanic and non-volcanic geosynclinal belts in Nevada and Newfoundland: Canadian Petroleum Geology Bull., v. 14, p. 579-599.* [17]

— *1967a, Stratigraphy and structure of northeastern Newfoundland bearing on drift in north Atlantic: Am. Assoc. Petroleum Geologist Bull., v. 51, p. 579-600.* [17, 39, 50]

— *1967b, Deer Lake to St. Barbe: Gander Conference, Newfoundland, field trips, Dept. of Geology, Columbia Univ., New York, 34 p.* [23, fig. 2]

— *1969, Long Point Formation and Taconic orogeny in western Newfoundland [abstr.]: Geol. Soc. America Abstracts with Programs for 1969, pt. 7, p. 120.* [25, 68, 72, fig. 2]

Kay, Marshall, and Colbert, E. H., *1965, Stratigraphy and life history: New York, John Wiley and Sons, 736 p.* [67]

Kindle, C. H., and Whittington, H. B., *1958, Stratigraphy of the Cow Head region, western Newfoundland: Geol. Soc. America Bull., v. 69, p. 315-342.* [25]

— *1959, Some stratigraphic problems of the Cow Head region in western Newfoundland: New York Acad. Sci. Trans., ser. 2, v. 22, no. 1, p. 7-18.* [25]

— *1965, New Cambrian and Ordovician fossil localities in western Newfoundland: Geol. Soc. America Bull., v. 76, p. 683-688.* [25]

Kümmel, H. B., and Weller, Stuart, *1902, The rocks of the Green Pond Mountain region: New Jersey Geol. Survey Ann. Rept. 1901, p. 1-51.* [47]

Lamarche, R. Y., *1962, Etude des conglomérats de la region d'Orford—Sherbrooke: Laval Univ., M. Sc. thesis, 70 p. (available by interlibrary loan).* [16]

Lapham, D. M., and Bassett, W. A., *1964, K-Ar dating of rocks and tectonic events in the piedmont of southeastern Pennsylvania: Geol. Soc. America Bull., v. 75, p. 661-668.* [17]

Leech, G. B., Lowden, J. A., Stockwell, C. H., and Wanless, R. K., *1963, Age determinations and geological studies: Canada Geol. Survey Paper 63-17, 140 p.* [16]

Lewis, J. V., Kümmel, H. B., and Johnson, M. E., *compilers, 1950, Geologic map of New Jersey: New Jersey Geol. Survey, 2 sheets, scale 1:250,000.* [fig. 8]

Lilly, H. D., *1967, Some notes on stratigraphy and structural style in central west Newfoundland (posthumous; edited by W. D. Brückner), in Neale, E.R.W., and Williams, Harold, eds., Collected papers on geology of the Atlantic region: Geol. Assoc. Canada Spec. Paper 4, p. 201-211.* [25]

Logan, W. E., *1863, Report on the geology of Canada: Canada Geol. Survey Rept. Progress 1843-1863, 983 p.* [33, 34, 68]

Long, L. E., *1969, Whole-rock Rb-Sr age of the Yonkers gneiss, Manhattan Prong: Geol. Soc. America Bull., v. 80, p. 2087-2090.* [40]

Long, L. E., and Kulp, J. L., *1962, Isotopic age study of the metamorphic history of the Manhattan and Reading Prongs: Geol. Soc. America Bull., v. 73, p. 969-995.* [9, 46, 47]

Lowden, J. A., Stockwell, C. H., Tipper, H. W., and Wanless, R. K., *1963, Age determinations and geological studies: Canada Geol. Survey Paper 62-17, 140 p.* [16]

Lyons, J. B., and Faul, Henry, *1968, Isotope chronology of the northern Appalachians, in Zen, E-an, White, W. S., Hadley, J. B., and Thompson, J. B., Jr., eds., Studies of Appalachian geology, northern and maritime: New York, Interscience Publishers, p. 305-318.* [9, 39]

MacLachlan, D. B., and Root, S. I., *1966, Comparative tectonics and stratigraphy of the Cumberland and Lebanon valleys: Field Conf. Pennsylvania Geologists 31st, Harrisburg, Pennsylvania, 1966, 90 p.* [29, 41, 46, fig. 2]

McBride, E. F., *1962, Flysch and associated beds of the Martinsburg Formation (Ordovician), central Appalachians: J. Sediment. Petrology, v. 32, p. 39-91.* [13]

McGerrigle, H. W., and Skidmore, W. B., *1968, Geological map of Gaspé Peninsula: Quebec Dept. Nat. Resources Map 1642, scale 1:253,440.* [36]

McKillop, J. H., *1963, Geology of the Corner Brook area, with emphasis on the carbonate deposits: Newfoundland Memorial Univ. Geology Rept. 1, 102 p.* [26, fig. 2]

Middleton, G. V., *1965, Paleocurrents in Normanskill graywackes north of Albany, New York: Geol. Soc. America Bull., v. 76, p. 841-844.* [13]

Morris, R. W., and Kay, Marshall, *1966, Ordovician graptolites from the middle Table Head Formation at Black Cove, near Port au Port, Newfoundland: J. Paleontology, v. 40, p. 1223-1229.* [25]

Naylor, R. S., *1968, Origin and regional relationships of the corerocks of the Oliverian domes, in Zen, E-an, White, W. S., Hadley, J. B., and Thompson, J. B., Jr., eds., Studies of Appalachian geology, northern and maritime: New York, Interscience Publishers, p. 231-240.* [8, 17, 39, 50]

Neale, E.R.W., Beland, Jacque, Potter, R. R., and Poole, W. H., *1961, A preliminary tectonic map of the Canadian Appalachian region based on age of folding: Canadian Mining and Metall. Bull., v. 64, p. 405-412.* [33]

Neale, E.R.W., and Kennedy, M. J., *1967, Relationship of the Fleur de Lys Group to younger groups of the Burlington Peninsula, Newfoundland, in Neale, E.R.W., and Williams, Harold, eds., Collected papers on geology of the Atlantic region: Geol. Assoc. Canada Spec. Paper 4, p. 139-169.* [17]

Osberg, P. H., *1965, Structural geology of the Knowlton-Richmond area, Quebec: Geol. Soc. America Bull., v. 76, p. 223-250.* [19, 35]

Osborne, F. F., *1956, Geology near Quebec City: Naturaliste Canadien, v. 83, p. 157-223.* [34, 35, 41, fig. 2, fig. 9]

Osborne, F. F., and Riva, John, *1966, Post-Levis beds of the Quebec Group at St. Apollinaire, Lotbinière Co., P. Q.: Naturaliste Canadien, v. 93, p. 145-151.* [5, 12, 34]

Pavlides, Louis, Boucot, A. J., and Skidmore, W. B., *1968, Stratigraphic evidence for the Taconic orogeny in the northern Appalachians, in Zen, E-an, White, W. S., Hadley, J. B., and Thompson, J. B., Jr., eds., Studies of Appalachian geology, northern and maritime: New York, Interscience Publishers, p. 61-82.* [8, 39]

Pearce, G. W., and Freda, G. N., *1969, Magnetization of the Perry Formation of New Brunswick, and the rotation of Newfoundland, discussion: Canadian J. Earth Sci., v. 6, p. 353-355.* [71]

Platt, L. B., Loring, R. B., and Stephens, G. C., *1969, Taconic events in the Hamburg 15' quadrangle, Pennsylvania [abstr.]: Geol. Soc. America Abstracts with Programs for 1969, pt. 1, p. 48-49.* [29, 41, fig. 2]

Poole, W. H., *1967, Tectonic evolution of Appalachian region of Canada, in Neale, E.R.W., and Williams, Harold, eds., Collected papers on geology of the Atlantic region: Geol. Assoc. Canada Spec. Paper 4, p. 9-51.* [16, 17, 40, 71]

Rasetti, Franco, *1946, Cambrian and Early Ordovician stratigraphy of the lower St. Lawrence valley: Geol. Soc. America Bull., v. 57, p. 687-706.* [33, 34]

Ratcliffe, N. M., *1965, Bedrock geology of the Great Barrington area, Massachusetts: Ph.D. thesis, Pa. State Univ., 213 p. (available on microfilm or by interlibrary loan).* [46]

— *1968, Stratigraphic and structural relations along the western border of the Cortlandt intrusives, in Finks, R. M., ed., Guidebook to field excursions: New York State Geol. Assoc., 40th Ann. Mtg., Flushing, New York, 1968, p. 197-220.* [46, 47]

— *1969a, Structural and stratigraphic relations along the Precambrian front in southwestern Massachusetts, in Bird, J. M., ed., Guidebook for field trips in New York, Massachusetts, and Vermont: New England Intercollegiate Geol. Conf., 61st Ann. Mtg., Albany, New York, 1969, p. 1-1 - 1-21.* [46]

— *1969b, Stratigraphy and deformational history of rocks of the Taconic Range near Great Barrington, Massachusetts, in Bird, J. M., ed., Guidebook for field trips in New York, Massachusetts, and Vermont: New England Intercollegiate Geol. Conf., 61st Ann. Mtg., Albany, New York, 1969, p. 2-1 - 2-23.* [46, 49]

— *1970, Ancient strike-slip fault tectonics in The Hudson Highlands and Manhattan Prong: New York Acad. Sci. Trans., ser. 2, v. 32, p. 1009-1021.* [9, 46]

Richardson, S. W., Gilbert, M. C., and Bell, P. M., *1969, Experimental determination of kyanite-andalusite and andalusite-sillimanite equilibria; the aluminum silicate triple point: Am. J. Sci., v. 267, p. 259-272.* [47]

Rickard, M. J., *1965, Taconic orogeny in the western Appalachians — Experimental application of microtextural studies to isotopic dating: Geol. Soc. America Bull., v. 76, p. 523-536.* [9, 46]

Riva, John, *1966, Upper Lévis graptolites from Cowansville, southern Quebec: J. Paleontology, v. 40, p. 220-221.* [35, fig. 2]

— *1968, Graptolite faunas from the Middle Ordovician of the Gaspé north shore: Naturaliste Canadien, v. 95, p. 1379-1400.* [16, 33, 36, 44, fig. 2]

— *1969a, Canajoharie and Utica in eastern North America [abstr.]: Geol. Soc. America Abstracts with Programs for 1969, pt. 1, p. 50.* [27]

— *1969b, Middle and Upper Ordovician graptolite faunas of St. Lawrence lowlands of Quebec, and of Anticosti Island, in Kay, Marshall, ed., North Atlantic — geology and continental drift: Am. Assoc. Petroleum Geologists Mem. 12, p. 513-556.* [27, 35, 67, fig. 2]

Robertson, W. A., Moy, J. L., and Park, J. K., *1968, Magnetization of the Perry Formation of New Brunswick, and the rotation of Newfoundland: Canadian J. Earth Sci., v. 5, p. 1175-1181.* [71]

— *1969, Magnetization of the Perry Formation of New Brunswick, and the rotation of Newfoundland, reply: Canadian J. Earth Sci., v. 6, p. 356.* [71]

Rodgers, John, *1963, Mechanics of Appalachian foreland folding in Pennsylvania and West Virginia: Am. Assoc. Petroleum Geologists Bull., v. 47, p. 1527-1536.* [7, 50]

— *1965, Long Point and Clam Bank Formations, western Newfoundland: Geol. Assoc. Canada Proc., v. 16, p. 83-94.* [23, 25, fig. 2]

— *1967, Chronology of tectonic movements in the Appalachian region of eastern North America: Am. J. Sci., v. 265, p. 408-427.* [9, 40]

— *1968, The eastern edge of the North American continent during the Cambrian and Early Ordovician, in Zen, E-an, White, W. S., Hadley, J. B., and Thompson, J. B., Jr., eds., Studies of Appalachian geology, northern and maritime: New York, Interscience Publishers, p. 141-150.* [31, 39, 40, 43]

Rodgers, John, and Neale, E.R.W., *1963, Possible "Taconic" klippen in western Newfoundland: Am. J. Sci., v. 261, p. 713-730.* [16, 17, 23, 25, 26, 45, fig. 2]

Roliff, W. A., *1968, Oil and gas exploration — Anticosti Island, Quebec: Geol. Assoc. Canada Proc., v. 19, p. 31-36.* [35, 68, 69]

Rosenfeld, J. L., *1968, Garnet rotations due to the major Paleozoic deformations in southeast Vermont, in Zen, E-an, White, W. S., Hadley, J. B., and Thompson, J. B., Jr., eds., Studies of Appalachian geology, northern and maritime: New York, Interscience Publishers, p. 185-202.* [8, 47]

Ruedemann, Rudolf, and Wilson, T. Y., *1936, Eastern New York Ordovician cherts: Geol. Soc. America Bull., v. 47, p. 1535-1586.* [39]

St-Julien, Pierre, *1961. Preliminary report on Lac Montjoie area, Sherbrooke, Richmond, and Stanstead counties: Quebec Dept. Nat. Resources Prelim. Rept. 464, 14 p.* [16]

— *1965, Orford-Sherbrooke area, Richmond, Sherbrooke, Shefford, Brome, Standstead and Compton Counties [Geologic map]: Quebec Dept. Nat. Resources Mineral Deposits Service Map 1619.* [16]

— *1967, Tectonics of part of the Appalachian region of southeastern Québec (southwest of the Chaudière River), in Clark, T. H., ed., Appalachian tectonics: Royal Soc. Canada Spec. Pub. 10, p. 41-47.* [16]

— *1968, Les "argiles-a-blocs" du sud-ouest des Appalaches du Québec: Naturaliste Canadien, v. 95, p. 1345-1356.* [16, 34, 35, 36, 41]

Schuchert, Charles, and Dunbar, C. O., *1934, Stratigraphy of western Newfoundland: Geol. Soc. America Mem. 1, 123 p.* [23, 25, fig. 2]

Schuchert, Charles, and Twenhofel, W. H., *1910, Ordovicic-Siluric section of the Mingan and Anticosti islands, Gulf of St. Lawrence: Geol. Soc. America Bull., v. 21, p. 677-716.* [67, 68]

Shaw, A. B., *1958, Stratigraphy and structure of the St. Albans area, northwestern Vermont: Geol. Soc. America Bull., v. 69, p. 519-568.* [19, fig. 2]

Sheridan, R. E., and Drake, C. L., *1968, Seaward extension of the Canadian Appalachians: Canadian Jour. Earth Sci., v. 5, p. 337-373.* [18, 71]

Sherwood, W. C., *1961, Structure of the Jacksonburg Formation in the Lehigh Valley, in Ryan, J. D., ed., Structure and stratigraphy of the Reading Hills and Lehigh valley in Northampton and Lehigh Counties, Pennsylvania: Field Conf. Pennsylvania Geologists 26th Ann. Mtg., Bethlehem, Pennsylvania, 1961, p. 40-43.* [46]

Sikander, A. H., and Fyson, W. K., *1969, Structural development of the Paleozoic rocks of western Gaspé, Quebec: Canadian J. Earth Sci., v. 6, p. 1113-1127.* [35]

Smith, C. H., *1958, Bay of Islands igneous complex, western Newfoundland: Canadian Geol. Survey Mem. 290, 132 p.* [fig. 2]

Snelgrove, A. L., *1931, Geology and ore deposits of Betts Cove—Tilt Cove area, Notre Dame Bay, Newfoundland: Canadian Mining and Metall. Bull., v. 24, no. 228, p. 477-519.* [17]

Stanley, R. S., *1964, The bedrock geology of the Collinsville quadrangle: Connecticut Geol. and Nat. History Survey, Quad. Rept. 16, 99 p.* [17]

Stäuble, Aloys, *1952, Les Cryptolothidés Québec: Naturaliste Canadien, v. 79, p. 285-319.* [34]

Stevens, R. K., *1967, Humber Arm area, Corner Brook: Gander Conference, Newfoundland, fieldtrips: Dept. of Geology, Columbia Univ., New York, 34 p.* [25, fig. 2]

— *1968, Taconic klippen of western Newfoundland: Canada Geol. Survey Paper 68-1, p. 8-10.* [23, 41, fig. 2]

— *1970, Cambro-Ordovician flysch sedimentation and tectonics in west Newfoundland and their possible bearing on a proto-Atlantic Ocean: Geol. Assoc. Canada Special Paper 7, p. 165-177.* [16, 23, 25]

Stevens, R. K., Church, W. R., and St-Julien, Pierre, *1969, Age of ultramafic rocks in the north-western Appalachians [abstr.]: Geol. Soc. America Abstracts with Programs for 1969, pt. 7, p. 215-216.* [16]

Stose, G. W., *1946, The Taconic sequence in Pennsylvania: Am. J. Sci., v. 244, p. 665-696.* [29, 31, fig. 2]

Thompson, J. B., Jr., *1967, Bedrock geology of the Pawlet quadrangle, Vermont; pt. 2, Eastern portion: Vermont Geol. Survey Bull. 30, p. 61-98.* [16, 27, 29]

Thompson, J. B., Jr., Robinson, Peter, Clifford, T. N., and Trask, N. J., *1968, Nappes and gneiss domes in west-central New England, in Zen, E-an, White, W. S., Hadley, J. B., and Thompson, J. B., Jr., eds., Studies of Appalachian geology, northern and maritime: New York, Interscience Publishers, p. 203-218.* [17, 39, 50]

Thompson, L.G.D., and Garland, G. D., *1957, Gravity measurements in Quebec (south of latitude 52°N.): Dominion Observatory Ottawa Pubs., v. 19, p. 111-167.* [36]

Tilton, G. R., Wetherill, G. W., Davis, G. L., and Hopson, C. A., *1958, Ages of minerals from the Baltimore gneiss near Baltimore, Maryland: Geol. Soc. America Bull., v. 69, p. 1469-1474.* [9, 39]

Tilton, G. R., Davis, G. L., Wetherill, G. W., Aldrich, L. T., and Jäger, Emilie, *1959, The ages of rocks and minerals: Carnegie Inst. Washington Year Book 58, p. 170-178.* [9]

Tilton, G. R., Wetherill, G. W., Davis, G. L., and Bass, M. N., *1960, 1,000-million-year-old minerals from the eastern United States and Canada: J. Geophys. Research, v. 65, p. 4173-4179.* [39]

Tuke, M. F., *1968, Autochthonous and allochthonous rocks in the Pistolet Bay area in northernmost Newfoundland: Canadian J. Earth Sci., v. 5, p. 501-513.* [12, 23, 26, 41, fig. 2]

Tuke, M. F., and Baird, D. M., *1967, Klippen in northern Newfoundland, in Clark, T. H., ed., Appalachian tectonics: Royal Soc. Canada Spec. Pub. 10, p. 3-9.* [26, fig. 2]

U.S. Geological Survey, *1964, Geological Survey Research 1964: U.S. Geol. Survey Prof. Paper 501-A, 367 p.* [31, fig. 2]

— *1966, Geological Survey Research 1966: U.S. Geol. Survey Prof. Paper 550-A, 385 p.* [33]

— *1968, Engineering geology of the northeastern corridor, Washington, D.C., to Boston, Massachusetts — Bedrock geology: U.S. Geol. Survey Misc. Geol. Inv. Map I-514-A.* [fig. 8]

— *1969, Geological Survey Research 1969: U.S. Geol. Survey Prof. Paper 650-A, 425 p.* [33]

U.S. Geological Survey and The American Association of Petroleum Geologists, *1962, Tectonic map of the United States, exclusive of Alaska and Hawaii (Prepared by a committee, G. V. Cohee, Chairman), 2 sheets, scale 1:2,500,000.* [71]

Weaver, D. F., *1967, A geological interpretation of the Bouguer anomaly field of Newfoundland: Dominion Observatory Ottawa Pubs., v. 35, p. 223-251.* [18, 26, 36, fig. 6]

Wetherill, G. W., Davis, G. L., and Lee-Hu, C., *1968, Rb-Sr measurements on whole rocks and separated minerals from the Baltimore gneiss, Maryland: Geol. Soc. America Bull., v. 79, p. 757-762.* [9, 39]

White, W. S., *Compiler, 1968, Geologic map of the northern and maritime Appalachian region, in Zen, E-an, White, W. S., Hadley, J. B., and Thompson, J. B., Jr., eds., Studies of Appalachian geology, northern and maritime: New York, Interscience Publishers, p. 453, scale 1:2,500,000.* [fig. 9]

Whittington, H. B., *1968, Zonation and correlation of Canadian and early Mohawkian series, in Zen, E-an, White, W. S., Hadley, J. B., and Thompson, J. B., Jr., eds., Studies of Appalachian geology, northern and maritime: New York, Interscience Publishers, p. 49-60.* [25]

Willard, Bradford, *1943, Ordovician clastic sedimentary rocks in Pennsylvania: Geol. Soc. America Bull., v. 54, p. 1067-1122.* [29, fig. 2]

Williams, Harold, *1967a, Silurian rocks of Newfoundland, in Neale, E.R.W., and Williams, Harold, eds., Collected papers on geology of the Atlantic region: Geol. Assoc. Canada Spec. Paper 4, p. 93-137.* [17]

— *Compiler, 1967b, Geology of the Island of Newfoundland: Canada Geol. Survey Map 1231A, scale 1:1,000,000.* [23, 26, fig. 2, fig. 6]

Williams, Harold, Kennedy, M. J., and Neale, E.R.W., *1970, The Hermitage flexure, the Cabot fault, and the disappearance of the Newfoundland central mobile belt: Geol. Soc. America Bull., v. 81, p. 1563-1568.* [26]

Williams, Harold, and Stevens, R. K., *1969, Geology of Belle Isle — northern extremity of the deformed Appalachian miogeosynclinal belt: Canadian J. Earth Sci., v. 6, p. 1145-1157.* [7, 71]

Wise, D. U., *1970, Multiple deformation, geosynclinal transitions and the Martic problem in Pennsylvania, in Fisher, G. W., Pettijohn, F. J., Reed, J. C., Jr., and Weaver, K. N., eds., Studies of Appalachian geology, central and southern: New York, Interscience Publishers, p. 317-333.* [31]

Woollard, G. P., and Joesting, H. R., *1964, Bouguer gravity anomaly map of the United States (exclusive of Alaska and Hawaii): U.S. Geol. Survey, 2 sheets, scale 1:2,500,000.* [18, fig. 3, fig. 7]

Zen, E-an, *1961, Stratigraphy and structure in the vicinity of the Taconic Range in west-central Vermont: Geol. Soc. America Bull., v. 72, p. 293-338.* [27, 41]

— *1967a, Time and space relationships of the Taconic allochthon and autochthon: Geol. Soc. America Spec. Paper 97, 107 p.* [11, 12, 27, 28, 29, 35, 39, 41, 44, 45, 49, fig. 2, fig. 7]

— *1967b, Stratigraphic-structural contrasts in thrust slices of the Taconic allochthon [abstr.]: Geol. Soc. America Abstracts for 1967, Geol. Soc. America Spec. Paper 115, p. 304-305.* [25, 45]

Zen, E-an, *1968, Nature of the Ordovician orogeny in the Taconic area, in Zen, E-an, White, W. S., Hadley, J. B., and Thompson, J. B. Jr., eds., Studies of Appalachian geology, northern and maritime: New York, Interscience Publishers, p. 129-140.* [20, 27, 28]

— *1969, Petrographic evidence for polymetamorphism in the western part of the northern Appalachians and a possible regional chronology [abstr.]: Geol. Soc. America Abstracts with Programs for 1969, pt. 7, p. 297-299.* [13, 27, 46]

Zen, E-an, and Ratcliffe, N. M., *1966, A possible breccia in southwestern Massachusetts and adjoining areas, and its bearing on the existence of the Taconic allochthon: U.S. Geol. Survey Prof. Paper 550-D, p. D39-D-46.* [45]

— *1971, Bedrock geologic map of the Egremont quadrangle and adjacent areas, Berkshire County, Massachusetts, and Columbia County, New York: U.S. Geol. Survey Misc. Geol. Inv. Map I-628, with 4-page explanatory text.* [45, 46]

Manuscript Received by The Society July 21, 1970
Revised Manuscript Received January 11, 1971
Publication Authorized by the Director, U.S. Geological Survey

Appendices

Appendix 1: Latitudinal Span of the Shelf Sequence and Its Implications

The lower Paleozoic shelf sequence on the west side of the Appalachian orogen extends today from the northwest tip of Newfoundland, with few gaps, to central Alabama, covering a span of latitude of about 20°. The carbonate rocks in the sequence are remarkable for their lithic similarity from one end of the belt to the other. What kind of paleoclimate could have given rise to this similarity? It is hard to believe that over an extended period of perhaps some 150 m.y. the world was persistently semitropical.

Perhaps the North American craton has been rotated since early Paleozoic time. Three virtual magnetic north pole positions have been compiled by Cox and Doell (1960, p. 699-700) for Middle and Upper Ordovician Appalachian rocks. Two of these pole positions from rocks of Trenton age make no sense when related to the shelf sequence. However, when the pole position determined by Collison and Runcorn (*in* Cox and Doell, 1960) from data on the Juniata Formation (Upper Ordovician), at long 153° E. and lat 20° N., is used as the north pole, the entire shelf sequence lies in a nearly due east-west belt at lat 10° S. This position could neatly explain the nature of the rocks, and would further place the similar shelf sequence of western United States in a nearly east-west belt 10° to 20° north of the equator.

Large-scale rotation of the North American craton, if substantiated, would demand large-scale movement of the continental crust and would make the idea of movement of continental plates and possible local fragmentation, suggested in this paper, more palatable.

Appendix 2: Tectonic Role of Anticosti Island during the Taconic Orogeny

Anticosti Island lies between the east tip of Gaspé Peninsula and the Great Northern Peninsula of Newfoundland (Fig. 1). The southeast half of the island is included in a gravity trough that coincides with Logan's Line (Goodacre and others, 1969). The role of this area during the Taconic orogeny is therefore of interest. Tectonically, the island has been regarded commonly as part of the foreland (*see*, for instance, Kay and Colbert, 1965, Fig. 10-16). However, the oldest rock unit exposed on the island is the Macasty Formation (Schuchert and Twenhofel, 1910), which is Middle Ordovician (Canajoharie and Utica; Riva, 1969b, p. 533). The Macasty is stratigraphically above the Mingan Limestone which is part of the "foreland" sequence exposed on the north shore of the St. Lawrence River, at Mingan Island, and elsewhere (Logan, 1863, p. 158-162; Dresser and Denis, 1944, map; Schuchert and Twenhofel, 1910, p. 688). The Mingan Limestone has been correlated with the post-tectonic Long Point Formation of Port-au-Port Peninsula of southwestern Newfoundland (Kay, 1969). Thus, the Macasty Formation is younger than the episode of gravity sliding in southwestern Newfoundland.

Recent drill-hole data from the northwest part of Anticosti Island (Roliff, 1968) show that the Macasty is underlain by carbonate rocks of the shelf sequence that reaches down without obvious stratigraphic break

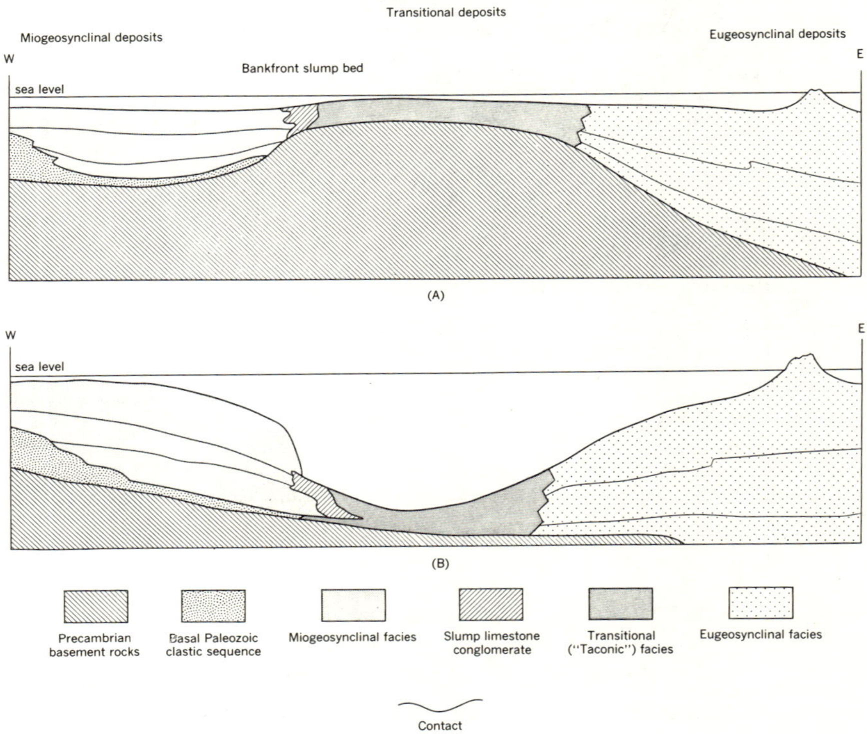

Figure 12. Schematic representation of paleogeographic relations for (A) a geanticline within a geosyncline and (B) a starved basin within a geosyncline. Basin-facies sediments from the top of the geanticline (A) or the central part of the starved basin (B) have later slid off as allochthons; in facies, these sediments are transitional between the typical eugeosynclinal rocks (right side) and the shelf rocks (left side). Stratigraphic convergence and thinning of sections exist in both sections. Direction of sediment transport, especially of turbidite and slump deposits, may be a useful criterion for choosing between (A) and (B). Significant vertical exaggeration.

into the Beekmantown, whose lithostratigraphy appears to differ from correlative units of western Newfoundland. There is no subsurface data in the southeastern part of the island, but the configuration of the top of the basement there (Roliff, 1968, p. 32) suggests stratigraphic or structural complications, or both. We cannot rule out the possibility that the southeast part of the island includes subsurface rocks deformed early during the Taconic orogeny, perhaps even buried gravity slide sheets.

Appendix 3: Paleogeographic Implications of the Transitional Facies

The paleogeographic significance of the transitional facies has been interpreted differently. Cady (1968b) envisioned that the Taconic rocks (and presumably all other similar rocks in the same belt of the Taconides) were deposited over a geanticline of shallow depth. He considered that the stratigraphic convergence of the typical thick eugeosynclinal facies westward into the transitional facies as evidence of a geanticline, where subsidence was less rapid.

The eugeosynclinal sequence is on the order of 6 km (20,000 ft) thicker than the synchronous transitional facies. As it seems unlikely that the water depth over the area of deposition of the transitional facies was 6 km deep at any time, more rapid subsidence of the eugeosyncline seems evident; to this extent Cady's (1968b) concept is clearly valid. However, the concept of an area of less rapid subsidence where the transitional facies was deposited does not imply an area of less deep water. Many features of the sediments in the transitional facies — chert, radiolaria, graptolites — indicate possible deep water; certainly the presence of the sequence of carbonate-clast conglomerates derived from the platform shows that the water depth was greater, possibly considerably greater, than that of the platform.

Thus, I envision the area of deposition of the transitional facies not as a geanticline but as the margin of the craton where it partakes of two features: relatively slow subsidence, shared by the area of the platform, but also deeper water than the platform area, receiving sediments of the shale-graywacke assemblage, in common with the eugeosyncline farther out to sea. The major difference between this area and the eugeosyncline is that the latter area subsided much more rapidly, and, in it, volcanic rocks and volcanogenic sediments were much more important sources. The area of deposition of the transitional facies may well have been a "starved basin" (Elam, 1960), where the thinness of the sedimentary column reflects both the distance from source of sediment supply and the sediment traps provided by the eugeosyncline and the platform (Fig. 12).

Tectonically, the area of deposition of the transitional facies may have been a geanticline or at least a half-geanticline; to an observer contemporary with the deposition of the rocks, however, a "starved basin" may be a more appropriate description.

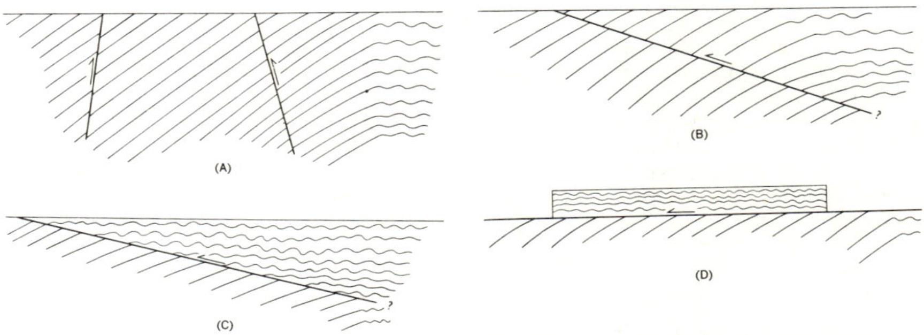

Figure 13. Schematic cross sections representing various geometric relations for autochthon and allochthon. (A), autochthon. (B) to (D), allochthon. (B) is a parautochthon, where the rocks presumably have not moved great distances and the indications are for local rooting. Generally, these rocks are of the same facies as the surrounding autochthon, if stratigraphically synchronous. (C) is a paraklippe, where the rocks presumably have moved a considerable distance, although complete detachment from the depositional site cannot be proven. Generally these rocks are of different facies from the surrounding autochthon, if stratigraphically synchronous. (D) is a klippe, where the rocks are demonstrably detached completely from the depositional site; they are now surrounded completely by the autochthon and separated from it by a movement surface. Different line patterns for the sections are intended to show different lithic facies.

Appendix 4: Autochthon, Neoautochthon, Allochthon, Parautochthon, Paraklippe, Klippe

The above terms are used in this paper in the following senses, which may not agree with those of some other authors.

Autochthon (Fig. 13a): Rocks that have not been moved appreciably from the original site of formation.

Neoautochthon: Autochthonous rocks deposited on an allochthon after movement has ceased.

Allochthon: Rocks that have been moved from the original site of formation. Under this general heading are recognized three sub-situations:

Parautochthon (Fig. 13b): Rocks that have been moved, generally on a fault surface of initially low dip. Complete severance from site of formation cannot be demonstrated and presumably did not happen. Amount of horizontal transport is commonly small (perhaps a few kilometers or less). If the rocks are sedimentary, the sedimentary facies of the transported and synchronous underlying rocks are commonly similar. If the rocks are metamorphic and the metamorphism is pre-movement, the contrast in metamorphic grade is also commonly small. Examples of parautochthonous rocks are the thrust slices of northwestern Vermont and the Precambrian rocks on the west front of the Berkshire massif.

Paraklippe (Fig. 13c): Rocks that have been moved, generally on a fault surface of initially low dip. Complete severance from the site of formation cannot be demonstrated. Amount of horizontal transport is commonly large (perhaps tens of kilometers). If the rocks are sedimentary, the facies of the transported and synchronous underlying rocks are commonly dissimilar. If the rocks are metamorphic and the metamorphism is pre-movement, the contrast in metamorphic grade also may be appreciable. Examples are the basin rocks immediately south of Logan's Line in Quebec, or rocks of the Pine Mountain plate (U.S. Geol. Surv. and Am. Assoc. Petrol. Geol., 1962) in the southern Appalachian region.

Klippe (Fig. 13d): Rocks that have been moved from the original site of formation, generally on a fault surface of initially low dip. The moved rocks are completely severed, by tectonic movement or by erosion, from the original site and are now in fault contact against the surrounding rocks on all sides. Rocks of the klippe commonly, though not necessarily, have been transported for long distances and are commonly distinct from those of the surrounding area, in sedimentary facies or in metamorphic grade. Examples are the Taconic klippe and the Humber Arm klippe.

There is no sharp distinction between paraklippe and parautochthon; a measure of judgment enters into the classification, and much depends on the particular frame of reference needed for the discussion of the subject matter. It is also clear that the distinction between a klippe and a paraklippe cannot always be made in the field. Moreover, a rock mass that is parautochthonous relative to the immediately underlying unit may be part of a klippe when a larger system is considered. In these situations, again, the application of a term is a matter of judgment and pertinence to the problem.

A "fault surface" may include a surface or zone of sliding of soft rock masses, and, thus, does not necessarily conform to the classic concept of a fault. In a sense, every clastic particle in a sedimentary rock is a klippe; indeed, there is probably every gradation between these clastic grains, pebbles, boulders, and major slices of the Taconic klippe, for instance. Again, judgment must enter into the choice of terminology to be used in each situation.

Appendix 5: Rotation of Newfoundland?

Black (1964) pointed out that data on orientation of virtual magnetic poles for the lower Paleozoic rocks suggest that Newfoundland had rotated counterclockwise by about 30° relative to the North American mainland, and that the rotation occurred between Early Devonian and Carboniferous time, presumably associated with the Acadian orogeny. The existing sparse data do not allow determination of the point about which the rotation occurred, nor the amount of translational slip that might have taken place.

Black's suggestion has received discussion (for example, Robertson and others, 1968, 1969; Pearce and Freda, 1969). Recent seismic studies in the vicinity of Newfoundland (Sheridan and Drake, 1968, p. 369) seem to rule out any major changes in crustal constitution in the pertinent area, as might be expected if the rotation had indeed occurred. Poole (1967, p. 44) and Williams and Stevens (1969, p. 1156) also discussed Black's suggestion and proposed geologic tests.

It is not clear whether Sheridan and Drake's (1968) data are fatal to Black's idea or all modifications of it. The idea does have certain geologic attractions. The tectonic trend of western Newfoundland is decidedly out of line with that of Gaspé Peninsula; reversal of a 30°

counterclockwise rotation would wipe out the discordance, and would also wipe out much of the peculiar sigmoid trace of the gravity ridge and trough in the Gulf of St. Lawrence. Such a reversal in rotation would also juxtapose southwestern Newfoundland and southeastern Anticosti Island, and make the similarity of the Long Point and Mingan Formations (Kay, 1969) readily explicable.